On

the Internet

Second edition

Thinking In Action

Series editors: Simon Critchley, New School University, New York, and Richard Kearney, University Gollege Dublin and Boston College

Thinking in Action is a major new series that takes philosophy to its public. Each book in the series is written by a major international philosopher or thinker, engages with an important contemporary topic, and is clearly and accessibly written. The series informs and sharpens debate on issues as wide ranging as the Internet, religion, the problem of immigration and refugees, and the way we think about science. Punchy, short and stimulating, **Thinking in Action** is an indispensable starting point for anyone who wants to think seriously about major issues confronting us today.

Praise for the series

'. . . allows a space for distinguished thinkers to write about their passions.'
The Philosophers' Magazine

'. . . deserve high praise.'
Boyd Tonkin, *The Independent* (UK)

'This is clearly an important series. I look forward to reading future volumes.'
Frank Kermode, author of *Shakespeare's Language*

'. . . both rigorous and accessible.'
Humanist News

'. . . the series looks superb.'
Quentin Skinner

'. . . an excellent and beautiful series.'
Ben Rogers, author of *A.J. Ayer: A Life*

'Routledge's Thinking in Action series is the theory junkie's answer to the eminently pocketable Penguin 60s series.'
Mute Magazine (UK)

'Routledge's new series, Thinking in Action, brings philosophers to our aid . . .'
The Evening Standard (UK)

'. . . a welcome new series by Routledge.'
Bulletin of Science, Technology and Society (Can)

'Routledge's innovative new Thinking in Action series takes the concept of philosophy a step further'
The Bookwatch

HUBERT L. DREYFUS

On
the Internet

Second edition

Routledge
Taylor & Francis Group

LONDON AND NEW YORK

First published 2001
by Routledge
2 Park Square, Milton Park, Abingdon, Oxon OX14 4RN

Simultaneously published in the USA and Canada
by Routledge
711 Third Avenue, New York, NY 10017 (8th Floor)

Routledge is an imprint of the Taylor & Francis Group, an informa business

© 2001, 2009 Hubert L. Dreyfus

Typeset in Joanna MT and DIN by
RefineCatch Limited, Bungay, Suffolk

British Library Cataloging in Publication Data
A catalogue record for this book is available from the British Library

Library of Congress Cataloguing in Publication Data
Dreyfus, Hubert L.
    On the internet / Hubert Dreyfus
        p.   cm. – (Thinking in action)
    Includes bibliographical references and index.
        1. Information technology – Social aspects.   2. Internet – Social aspects.   3. Social
isolation.   I. Title.   II. Series.

    HM851 .D74 2001
    303.48'33 – dc21   00–046010

ISBN10: 0–415–77516–7 (pbk)
ISBN10: 0–203–88793–X (ebk)
ISBN13: 978–0–415–77516–8 (pbk)
ISBN13: 978–0–203–88793–6 (ebk)

For Geneviève,
surfing survivor and Website designer,
who has mastered the worst and best of the Internet.

"Body am I, and soul" – thus speaks the child. And why should one not speak like children?

But the awakened and knowing say: body am I entirely, and nothing else; and soul is only a word for something about the body.

Friedrich Nietzsche, **Thus Spake Zarathustra**,

trans. W. Kaufmann, New York: Viking Press, 1966, p. 34

The body is our general medium for having a world. Sometimes it is restricted to the actions necessary for the conservation of life, and accordingly it posits around us a biological world; at other times, elaborating upon these primary actions and moving from their literal to a figurative meaning, it manifests through them a core of new significance: this is true of motor skills such as dancing. Sometimes, finally, the meaning aimed at cannot be achieved by the body's natural means; it must then build itself an instrument, and it projects thereby around itself a cultural world.

Maurice Merleau-Ponty, **Phenomenology of Perception**,

trans. C. Smith, London: Routledge & Kegan Paul, 1962, p. 146

Acknowledgements

I'm indebted to many people for their invaluable help: to Nat Goldhaber for espousing the virtues of the Net's disembodiment; to Stuart Dreyfus for teaching me all I know about skill acquisition; to Hal Varian and Gordon Rios for their patient explanations of how searching on the Net actually works; to Arun Tripathi for forwarding to me more material on the Internet than anyone could ever have time to read; to Kenneth Goldberg, Charles Spinosa, Sean Kelly, Béatrice Han, Corbin Collins, Mark Wrathall and Terry Winograd for tough objections and help in answering them; and to Jos de Mul and his seminar for working through the whole manuscript and making many helpful critical comments; to Geneviève Dreyfus for teaching me to use the Net and for preparing the final manuscript; and especially to David Blair, whose sophisticated Wittgensteinian understanding of document retrieval not only helped me understand the problems of search on the Internet, but also helped me see how these problems fit with my own Merleau-Pontian sense of the limitations of life in cyberspace. Finally, thanks to Philip Rosedale, CEO and Founder of Linden Lab, for filling me in on the latest work there.

The extract from Bob Dylan's "Highway 61 Revisited" on p. 72 is reprinted by kind permission of Sony/ATV Music Publishing (UK) Ltd.

Preface to the Second Edition

The world of the Internet is changing so rapidly that to bring *On the Internet* up to date I had to make some serious changes, and add a whole new chapter.

The most radical change is in Chapter One. There I endorsed the current pessimism concerning the possibility of successfully searching billions of meaningless hypertext websites. Now, that pessimism has turned to optimism thanks to Google – a program which was just a proposed PhD thesis ten years ago when I was finishing the first edition. So I've cut out all the gloomy predictions and added an explanation of how Google works, and how Wikipedia, also new in this decade, is gaining followers using the old meaning-based ordering of information.

Likewise, most of Chapter Two predicting the failure of disembodied distance learning and ridiculing the enthusiasts who claimed that, thanks to the Internet, an Ivy League education would be available to everyone on the planet and that universities as we know them would disappear had to be scrapped. It is now clear that distance learning has failed. The major universities have given up on it and consider their investments of hundreds of thousands of dollars as sunk costs.

Something altogether new, however, has come along in the past decade to provide an unexpected and original form of distance learning – the podcast which doesn't try to replace embodied classroom teaching but which offers an exciting opening on first-class courses nonetheless. I include as an Appendix an article from the *LA Times* describing the development of iTunesU and my involvement in it.

People from all over the world have sent me e-mails saying they wish they were able to meet with others who are listening to my podcasts and talk with them and with me, so I've experimented with teaching a virtual discussion section in *Second Life*. I describe the results in Chapter Five.

Chapter Five is a totally new chapter about three-dimensional interactive virtual worlds. *Second Life* is the most prominent example of how one can create and control a virtual body in a virtual world, so I focus on which aspects of embodiment can be captured in *Second Life* and which cannot, and how this affects what sort of meaningful lives are and are not possible on the Internet.

Introduction

The Internet is not just a new technological innovation; it is a new *type* of technological innovation; one that brings out the very essence of technology. Up to now, technological innovators have generally produced devices that served needs that were already recognized, and then discovered some unexpected side effects. So Alexander Graham Bell thought the telephone would be useful for communication in business but would not be accepted into people's homes, let alone intrude as they walked down the street. Likewise, Henry Ford thought of the automobile as giving people cheap reliable, individualized transportation, but he did not imagine it would destroy the inner cities and liberate adolescent sex. The Net is different. It was originally intended for communication between scientists, but now *that* is a side effect. We have come to realize that the Net is too gigantic and protean for us to think of it as a device for satisfying *any* specific need, and each new use it affords is a surprise. If the essence of technology is to make *everything* accessible and optimizable, then the Internet is the perfect technological device. It is the culmination of the same tendency to make everything as flexible as possible that has led us to digitalize and interconnect as much of reality as we can.[2] What the Web will allow us to do is literally unlimited. This pure flexibility naturally leads people to vie for outrageous predictions as to what the Net will become. We are told that, given its new way of linking and accessing information, the Internet will bring a new era of economic prosperity, lead to the development of intelligent search engines that will deliver to us just the information we desire, solve the problems of mass education, put us in touch with all of actual reality, enable us to explore virtual worlds that enable us to have even more flexible identities than we have in the real world and thereby add new dimensions of meaning to our lives.

Unfortunately, work in areas where a new and more fulfilling form of life has been promised has produced a great deal of talk but few happy results.[3] In fact, researchers at Carnegie-Mellon University were surprised to find that, when people were given access to the World Wide Web, they found themselves feeling isolated and depressed. *The New York Times* reports:

> The results of the $1.5 million project ran completely contrary to expectations of the social scientists who designed it and to many of the organizations that financed the study. . . .
> "We were shocked by the findings, because they are counterintuitive to what we know about how socially the Internet is being used," said Robert Kraut, a social psychology professor at Carnegie Mellon's Human Computer Interaction Institute. "We are not talking here about the extremes. These were normal adults and their families, and

on average, for those who used the Internet most, things got worse."[4]

Other researchers sum up their findings as follows:

> This research examined the social and psychological impact of the Internet on 169 people in seventy-three households during their first one to two years on-line. . . . In this sample, the Internet was used extensively for communication. Nonetheless, greater use of the Internet was associated with declines in participants' communication with family members in the household, declines in the size of their social circle, and increases in their depression and loneliness.[5]

The authors conclude that what is missing is people's actual embodied presence to each other:

> On-line friendships are likely to be more limited than friendships supported by physical proximity. . . . Because on-line friends are not embedded in the same day-to-day environment, they will be less likely to understand the context for conversation, making discussion more difficult and rendering support less applicable. Even strong ties maintained at a distance through electronic communication are likely to be different in kind and perhaps diminished in strength compared with strong ties supported by physical proximity. The interpersonal communication applications currently prevalent on the Internet are either neutral toward strong ties or tend to undercut rather than promote them.[6]

This surprising discovery shows that the Internet user's disembodiment has profound and unexpected effects. Presumably, it affects people in ways that are different from the way most tools do because it can become the main way its users relate to the rest of the world. Given these surprises and disappointments, we would naturally like to know what are the benefits and the dangers of living our lives on-line? Only then might we hope to have a glimmer concerning what the Net can become and what we will become in the process of living through it.

According to the most extreme Net enthusiasts, the long-range promise of the Net is that each of us will be able to transcend the limits imposed on us by our body. As John Perry Barlow, one of the foremost proponents of this vision, puts it, the electronic frontier is "a world that is both everywhere and nowhere, but it is not where bodies live".[7] By our body, such visionaries seem to mean not only our physical body with its front and back, arms and legs, and ability to move around in the world, but also our moods that make things matter to us, our location in a particular context where we have to cope with real things and people, and the many ways we are exposed to disappointment and failure as well as to injury and death. In short, by embodiment they include all aspects of our finitude and vulnerability. In the rest of this book, I will understand the body in these broad terms.

Yeats lamented that his soul was "fastened to a dying animal" ("Sailing to Byzantium", in *The Tower*, 1928) and it is easy to see the attraction of completing human evolution by leaving behind the animal bodies in which our linguistic and cultural identities are now imprisoned. Who wouldn't wish to become a disembodied being who could be anywhere in the universe and make backup copies of himself to avoid injury and death? Not only Web visionaries would be delighted to be free from deformities, depression, sickness, old age, and death. This is the promise offered us by computer-inspired futurists such as Hans Moravec[8] and Ray Kurzweil.[9] It is typified on the Net (where else?) by such international groups as the Extropians, whose leader, Max More, is quoted in the epigraph to this Introduction. But even more down-to-earth gurus subscribe to the dream that we are entering a new level of civilization. According to industry consultant Esther Dyson, "Cyberspace is the land of knowledge, and the exploration of that land can be a civilization's truest, highest calling."[10]

Leaving the body behind would have pleased Plato, who subscribed to the saying that the body was the tomb of the soul[11] and followed Socrates in claiming that it should be a human being's highest goal to "die to his body" and become a pure mind. As Socrates put it: "In despising the body and avoiding it, and endeavoring to become independent — the philosopher's soul is ahead of all the rest."[12] But that makes it surprising that the Extropians claim to be following Nietzsche, not Plato, when they say we should transcend our humanity.

In fact, Nietzsche's anti-Platonic view of the body is in the very book about the overman the Extropians love to quote. In a section called "On the Despisers of the Body" Nietzsche has Zarathustra say, as if in direct response to the Extropians: "I shall not go your way, O despisers of the body! You are no bridge to the overman!"[13] And he continues:

> "I," you say, and are proud of the word. But greater is that in which you do not wish to have faith – your body and its great reason: that does not say "I," but does "I." . . . Behind your thoughts and feelings, my brother, there stands a mighty ruler, an unknown sage – whose name is self. In your body he dwells; he is your body.[14]

Nietzsche thought that the most important thing about human beings was not their intellectual capacities but the emotional and intuitive capacities of their body. In his relentless battle against Platonism and Christianity, even in its most hidden forms in science and technology, Nietzsche, indeed, looked forward to our transcending our human limitations and becoming overmen, but by that he meant that human beings, rather than continuing to deny death and finitude, would finally have the strength to affirm their bodies and their mortality.

So the issue we have to face is: can we get along without our bodies? Is the body just a remnant of our descent from the animal – a limitation on our freedom which the human race is now positioned to outgrow, as the Extropians claim – or does the body play a crucial role even in our spiritual and intellectual life, as Nietzsche contends? If Nietzsche is right, the Net's supposed greatest advantage, freedom from the limits imposed by our bodies, is, ironically, its Achilles' heel.

As a philosopher, I'm not going to become involved in condemning some specific uses of the Internet and praising others. My question is a more speculative one: what if the Net became central in our lives? What if it becomes, as the developers of *Second Life* hope it will become, what Joseph Nye, Dean of Harvard University's Kennedy School of Government, calls an "irresistible alternative culture"? What if the Internet gave us access to a virtual second life? To the extent that we came to live a large part of our lives in cyberspace, would we become super- or infra-human?

In seeking an answer, we should remain open to the possibility that, when we enter cyberspace and leave behind our emotional, intuitive, situated, vulnerable, embodied selves, and thereby gain a remarkable new freedom never before available to human beings, we might, at the same time, necessarily lose some of our crucial capacities: our ability to make sense of things so as to distinguish the relevant from the irrelevant, our sense of the seriousness of success and failure that is necessary for learning, and our need to get a maximum grip on the world that gives us our sense of the reality of things. Furthermore, we would be tempted to avoid the risk of genuine commitment, and so lose our sense of what gives meaning to our lives. Indeed, in what follows, I hope to show that, if our body goes, and we live, for example, through avatars (virtual bodies) as in *Second Life*, we will largely lose our sense of relevance, our ability to acquire skills, our sense of resistant reality, our ability to make maximally meaningful commitments, and the embodied moods that give life serious meaning. If that is the trade-off, the prospect of living our lives in and through the Web may not be so attractive after all.

SUMMARY

Chapter One. The limitations of hyperlinks. The hope for intelligent information retrieval, and the failure of Artificial Intelligence (AI). How the actual shape and movement of our bodies play a crucial role in our making sense of our world, so that loss of embodiment would lead to loss of the ability to recognize relevance.

Chapter Two. The myth of distance learning. The importance of mattering for teaching and learning. Apprenticeship and the need for imitation. Without involvement and presence we cannot acquire skills.

Chapter Three. Telepresence as absence. The body as source of our grip on reality. How the loss of embodied coping in telepresence would lead to the loss of a sense of the reality of people and things.

Chapter Four. Anonymity and nihilism. Maximal meaning in our lives requires genuine commitment and real commitment requires real risks. The anonymity and safety of life on the Web necessarily lacks such serious meaning.

Chapter Five. Moods and mattering. Unless people in virtual worlds can experience moods, they will not have have memorable, focal experiences, let alone maximally meaningful ones.

One

The AI Problem, as it's called – of making machines behave close enough to how humans behave intelligently – . . . has not been solved. Moreover, there is nothing on the horizon that says, I see some light. Words like "artificial intelligence," "intelligent agents," "servants" – all these hyped words we hear in the press – are restatements of the mess and the problem we're in.

We would love to have a machine that could go and search the Web, and our personal stores, knowing our preferences, and knowing what we mean when we say something. But we just don't have anything at that level.

Michael Dertouzos, Director, Laboratory for Computer Science, MIT[1]

Successful retrieval of information is the primary goal of most Web users. According to a Pew Foundation report: "Search engines are highly popular among Internet users. Searching the Internet is one of the earliest activities people try when they first start using the Internet, and most users quickly feel comfortable with the act of searching . . . 84% of internet users have used search engines and, on a given day, 56% of those online use a search engine."[2] As everyone who has searched on the Web knows, the power of search engines has changed dramatically in the past decade. To understand the current situation and to anticipate future developments we need to understand the problems involved in providing quick and reliable searches, how it was done a decade ago and how it is done now.

When I finished the manuscript of this book in 1999, the people whose judgment I trusted were deeply pessimistic about the future of information retrieval on the World Wide Web. The issues they raised are still relevant, although, as we shall soon see, their pessimism is not. In this second edition I will retain a shortened and lightly edited version of my opening remarks from the first edition up to the point where the then current understanding of the problem of search became history, and the attitude of the reliable researchers changed almost overnight. Then, in the new material that makes up the second half of this chapter, I'll explain what is possible now, how it became possible, and, based on these new developments, I'll predict where search is going from here.

In 1999 I wrote:

The Web is vast and growing exuberantly. At a recent count, it had over a billion pages and it continues to grow at the rate of at least a million pages a day.[3] (It is characteristic of the Web that these statistics, as you read them, are already far out of date.) There is an amazing amount of useful information on the Web but it is getting harder and harder to find. The problem arises from the way information is organized (or, better, disorganized) on the Web. The way the Web works, each element of this welter of information is linked to many other elements by hyperlinks. Such links can link any element of information to any other element for any reason that happens to occur to whoever is making the link. No authority or agreed-upon catalogue system constrains the linker's associations.[4]

Hyperlinks have not been introduced because they are more useful for retrieving relevant information than the old systematic ordering. Rather, they are the natural way to use the speed and processing power of computers to relate a vast amount of information without needing to understand it or impose any authoritarian or even generally accepted structure on it. But, when everything can be linked to everything else without regard for purpose or meaning, the vast size of the Web and the arbitrariness of the links make it extremely difficult for people desiring specific information to find the information they seek.

The traditional way of ordering information depends on someone – a zoologist, a librarian, a philosopher – having worked out a classification scheme according to the meanings of the terms involved and the interests of the users.[5] People can then enter new information into this classification scheme on the basis of what they understand to be the meaning of the categories and of the new information. If one wants to use the information, one has to depend on those who developed the classifications to have organized the information on the basis of its meaning so that users can find the information that is relevant given their interests.

Since Aristotle, we have been accustomed to organize information in a hierarchy of broader and broader classes, each including the narrower ones beneath it. So we descend from things, to living things, to animals, to mammals, to dogs, to collies, to Lassie. When information is organized in such a vertical database, the user can follow out the meaningful links, but the user is forced to commit to a certain class of information before he can view more specific data that fall under that class. For example, I have to commit to an interest in animals before I can find out what I want to know about tortoises; and once having made that commitment to the animal line in the database, I can't then examine the data on problems of infinity without backtracking through the commitments I have made.

When information is organized horizontally by hyperlinks, however, as it is on the Web, instead of the relation between a class and its members, the organizing principle is simply the inter-connectedness of all elements. There are no hierarchies; everything is linked to everything else on a single level, and meaning is irrelevant. Thus hyperlinks allow the user to move directly from one data entry to any other, as long as they are related in at least some tenuous fashion. The whole of the Web lies only a few links away from any page. With a hyper-linked database, the user is encouraged to traverse a vast network of information, all of which is equally accessible and none of which is privileged. So, for instance, among the sites that contain information on tortoises suggested to me by my browser, I might click on the one called "Tortoises – compared to hares", and be transported instantly to an entry on Zeno's paradox.

We can focus the old and new ways of organizing and retrieving information, and see the attraction of each, by contrasting the old library culture and the new kind of libraries made possible by hyperlinks. Table 1 contrasts a meaning-driven, semantic structuring of information with a formal, syntactic structuring, where meaning plays no role.

Clearly, the user of a hyper-connected library would no longer be a modern subject with a fixed identity who desires a more complete and reliable model of the world,[6] but rather a postmodern, protean being ready to be opened up to ever new horizons. Such a new being is not interested in collecting what is significant but in connecting to as wide a web of information as possible.

Web surfers embrace proliferating information as a contribution to a new form of life in which surprise and wonder are more important than meaning and usefulness. This approach appeals especially to those who like the idea of rejecting hierarchy and authority and who don't have to worry about the practical problem of finding relevant information. So postmodern theorists and artists embrace hyperlinks as a

OLD LIBRARY CULTURE	HYPERLINKED CULTURE
Classification	**Diversification**
a. stable	a. flexible
b. hierarchically organized	b. single-level
c. defined by specific interests	c. allowing all possible associations
Careful selection	**Access to everything**
a. quality of editions	a. inclusiveness of editions
b. authenticity of the text	b. availability of texts
c. eliminate old material	c. save everything
Permanent collections	**Dynamic collections**
a. preservation of a fixed text	a. intertextual evolution
b. interested browsing	b. playful surfing

Table 1: Opposition between old and new systems of information retrieval

way of freeing us from anonymous specialists organizing our databases and deciding for us what is relevant to what. Quantity of connections is valued above the quality of these connections. The idea has an all-American democratic ring. As Fareed Zakaria, the managing editor of *Foreign Affairs*, observes: "The Internet is profoundly disrespectful of tradition, established order, and hierarchy, and that is very American."[7]

Those who want to use the available data, however, have to find the information that is meaningful and relevant to them given their current concerns. But, given that in a hyperlinked database anything may be linked to anything else, this is a very challenging task. Since hyperlinks are made for all sorts of reasons and since there is only one basic type of link, the searcher cannot use the meaning of the links to arrive at the information he is seeking. The problem is that, as far as meaning is concerned, all hyperlinks are alike. As one researcher puts it, the retrieval job is worse than looking for a needle in a haystack; it's like looking for a specific needle in a needle stack. Given the lack of any semantic content determining the connections, it looks like any means for searching the Web must be a formal, syntactic technique called data mining that tracks statistical relations such as frequency between meaningless data.

The difficulty of using meaningless mechanical operations to retrieve meaningful information did not await the arrival of the Net. It arises whenever anyone seeks to retrieve information relevant to a specific purpose from a database not organized to serve that particular purpose. In a typical case, researchers may be looking for published papers on a topic they are interested in, but the mere words in the titles of the papers do not enable a search engine to return just those documents or websites that meet a specific searcher's needs.

To understand the problem it helps to distinguish Data Retrieval (DR) from Information Retrieval (IR). David Blair, Professor of Computer and Information Systems at the University of Michigan,[8] explains the difference:

> Data Base Management Systems have revolutionized the management and retrieval of data – we can call directory assistance and get the phone number of just about anyone anywhere in the US or Canada; we can walk to an ATM in a city far away from our home town and withdraw cash from our home bank account; we can go to a ticket office in Michigan and buy a reserved seat for a play in San Francisco; etc. All of this is possible, in part, because of the large-scale, reliable database management systems that have been developed over the last 35 years.

> Data retrieval operates on entities like "names,"
> "addresses," "phone numbers," "account balances,"
> "social security numbers," – all items that typically have
> clear, unambiguous references. But although some of the
> representations of documents have clear senses and
> references – like the author or title of a document – many IR
> searches are not based on authors or titles, but are interested
> in the "intellectual content" of the documents (e.g., "Get
> me any reports that analyse Central European investment
> prospects in service industries"). Descriptions of intellectual
> content are almost never determinate, and on large retrieval
> systems, especially the WWW, subject descriptions are
> usually hopelessly imprecise/indeterminate for all but the
> most general searching.[9]

So searching for a known URL on the WWW is simple and easy; it has the precision and directedness of data retrieval. But searching for a Web page with specific intellectual content using Web search engines can be very difficult, sometimes impossible.

The difference between Data Retrieval and Document Retrieval can be summed up as shown in Table 2.

Before the advent of the Web and Web search engines, the attempted solution to the document retrieval problem was to have human beings – that is, indexers who understood the documents – help describe their contents so that they might be retrieved by those who wanted them. But there simply aren't enough cataloguers to index the Web – it's too large and it's growing too fast.

The early search engines simply created an index of words associated with a list of documents that contained them, with scoring based on whether or not the word was in the title,

DATA RETRIEVAL	DOCUMENT RETRIEVAL
1. Direct ("I want to know X")	1. Indirect ("I want to know about X")
2. Necessary relation between a request and a satisfactory answer	2. Probabilistic relation between a request and a satisfactory document
3. Criterion of success = correctness	3. Criterion of success = utility
4. Scaling up is not a major problem	4. Scaling up is a major problem

Table 2: The differences between data retrieval and document retrieval

body, abstract, etc. Researchers generally agree, however, that these techniques have only about a 10 per cent chance of retrieving a useful document for a given query. And it's clear that the ideal would be a syntactic search approach that would be as fast as a computer but would have the added advantage of understanding importance and relevance in human terms. This sort of desperation could lead to every new attempt to use Artificial Intelligence and natural language understanding to guide search.

Since the 1960s, AI researchers had been seeking to solve the problem of getting computers, which are syntactic engines sensitive only to the form or shape of their input, to behave like human beings who are sensitive to semantics or meaning. So, naturally, researchers trying to develop search techniques for the Web turned to AI for help in programming computers to find just those Web pages whose relevance would be recognized by a human being conducting a search.

In the 1960s AI researchers had been optimistic. They felt confident that they could represent the few million explicit facts about the world people knew and then use rules for finding which facts were relevant in any given situation. But in the late 1970s and early 1980s AI researchers reluctantly came to recognize that, in order to produce artificial intelligence, they would have to make explicit and organize the commonsense knowledge people share, and that was a huge task.[10]

The most famous proponent of this approach is Douglas Lenat.[11] Lenat understood that our commonsense knowledge is not the sort of knowledge found in encyclopedias, but, rather, is the sort of knowledge taken for granted by those writing articles in encyclopedias. Such background knowledge is so obvious to us that we hardly ever notice it. Lenat points out that to understand an article about George Washington, for example, we may need to know such facts as that, when he was in the Capitol, so was his left foot, and that, when he died, he stayed dead. So, in 1985, Lenat proposed that, over the next ten years, he would capture this common sense by building "a single intelligent agent whose knowledge base contains . . . millions of entries".[12]

Lenat has now spent fifteen years and at least fifteen million dollars developing CYC, a commonsense knowledge database, in the attempt to enable computers to understand commonsense concerns such as requests for information. To demonstrate the use of CYC, Lenat developed a photograph retrieval system as an example of how commonsense knowledge plays an essential role in information retrieval. The system is supposed to retrieve on-line images by caption. Instead of a billion images as one might find on the Web;

Lenat starts modestly with twenty pictures. A Stanford professor describes his experience with the system as follows:

> The CYC demo was done with 20 images. The request, "Someone relaxing", yielded one image, 3 men in beachwear holding surfboards. CYC found this image by making a connection between relaxing and previously entered attributes of the image. But even for 20 pictures the system does not work very well.[13]

In so far as this system works at all, it works only because CYC programmers have made explicit as *knowledge* some of the *understanding* we have of relaxation, exercise, effort, and so forth just by having bodies. But most of our understanding of what it's like to be embodied is so pervasive and action-oriented that there is every reason to doubt that it could be made explicit and entered into a database in a disembodied computer.

That, of course, is not a problem for us in our everyday lives. We can find out the answers to questions involving the body by using our body or imagining what it would be like to be doing such and such. So, for example, we understand that pushups are not relaxing, simply by imagining carrying out the activity. But, a picture of someone doing pushups would have to be labelled for CYC by a human programmer as someone making an effort. Only then could CYC "deduce" that the person was not relaxing.

In general, by having bodies we can generate as needed an indefinitely large number of facts about our bodies, so many that we do not and could not store them all as explicit knowledge. But CYC does not have a body, so, as we have seen, it has to be given all the facts about the body that it needs to retrieve information from its database. Moreover, CYC would still not understand how to use the facts it did know to answer some new question involving the body. For example, if one asked CYC if people can chew gum and whistle at the same time, it would have no idea of the answer even if it knew a lot of facts about chewing and whistling, until an embodied human being imagined trying to do it, and then added the answer to CYC's database. But the number of such facts about the body that one would need to make explicit and store because they might be relevant to some request is endless. Happily, by having a body we dispense with the need to store any such facts.

When Lenat embarked on his project fifteen years ago, he claimed that in ten years CYC would be able to read articles in the newspaper and catalogue the new facts it found there in its database without human help. This is the dream of those who expect artificial intelligent agents to find and deliver to each person the information he or she is interested in. But, as Michael Dertouzos makes clear in the epigraph at the head of this chapter, this breakthrough has not occurred. The moral is, as Don Swanson, former Dean of the Library School at the University of Chicago, points out, that "machines cannot recognize meaning and so cannot duplicate what human judgment . . . can bring to the process of indexing and classifying documents".[14]

The failure of AI projects such as Lenat's should call our attention to how important our bodies are in making sense of the world. Indeed, our form of life is organized by and for beings embodied like us: creatures with bodies that have hands and feet, insides and outsides; that have to balance in a gravitational field; that move forward more easily than backwards; that get tired; that have to approach objects by traversing the intervening space, overcoming obstacles as they proceed, etc. Our embodied concerns so pervade our world that we don't notice the way our body enables us to make sense of it.[15] We would only notice it by experiencing our disorientation if we were transported to an alien world set up by creatures with radically different – say, spherical or gaseous – bodies, or by observing the helpless confusion of such alien creatures brought into our world.

It would obviously be a great help if we could use our embodied sense of what is relevant for beings with bodies and interests like ours as a background whenever we searched the databases and websites of the world for relevant information. But, as Lenat's failure to achieve his goal of making explicit our commonsense knowledge has shown, there is no reason to hope we can formalize the understanding we have by virtue of being embodied. Indeed, the hope that Artificial Intelligence could solve the relevance problem has now been largely abandoned. There is a vast and ever-growing amount of information out there, and it looks like our only access to it will have to be through computers that don't have bodies, don't share our world, and so don't understand the meaning of our documents and websites.

If we leave our embodied commonsense understanding of the world aside, as using computers seems to force us to do, we have to do things the computer's way and try to locate relevant information by replacing semantics with correlations between formal squiggles. So there is a whole information retrieval industry devoted to developing Web crawlers and search engines that attempt to approximate a human being's sense of relevance by using only the statistical corrections of the meaningless symbols available to a computer.

But, given the immense size of the Net, it is estimated that syntactic search engines can find at most 2 per cent of the relevant sites. Indeed, faith in incremental progress towards being able to find just the information one needs only makes sense if there is an agreed upon taxonomy, like that of Aristotle or the Dewey decimal system, that captures the way the world is divided up. But in a world of hyperlinks, there can be no such saving metaphysical solution.

Don Swanson sums up the point succinctly:

> Consistently effective fully automatic indexing and retrieval is not possible. Our relevance judgments . . . entail knowing who we are, what we are, the kind of world we live in, and why we want what we seek. It is hardly imaginable that a mechanism . . . could acquire such self-knowledge, be given it, or do the job without it.[16]

Such was the generally shared pessimism about the future of search on the Web when I handed in the manuscript of this book in 1999. The fact that the Web was huge and growing rapidly was legitimate ground for discouragement or for a desperate hope that some new form of disembodied AI would make possible some sort of semantic search.[17] But reasonable resignation like Swansons's notwithstanding, a radical new approach to syntactic search was already on the drawing board. Surprisingly, it was an approach for which the growing size of the Web was not a cause for despair but an occasion for optimism.

Terry Winograd, an AI pioneer at the Massachusetts Institute of Technology, gave up on AI and moved to Stanford University, where he started teaching Martin Heidegger in his Computer Science courses. In 1965 he had served as dissertation adviser to a Stanford student named Larry Page who was working on a research project involving Web search. Winograd, who understands the limits both of mere statistical correlations, on the one hand, and the failure of AI, on the other, opened a space for his graduate student to develop a search procedure that showed that one could use the billions of meaningless hyperlinks on the Web, not by indexing meaningless keywords nor by understanding their meaningful content but by mining the "importance" of Web pages to human beings who are searching with some particular interest in mind. The breakthrough is to see that, while the horizontal syntactic hyperlinks can link anything to anything, the fact that people seeking relevant information have clicked on certain sites and not others can be mined for meaning. The idea was "to build a practical large-scale system which [could] exploit the additional information present in hypertext . . . to effectively deal with uncontrolled hypertext collections where anyone can publish anything they want".[18]

Winograd and his students report their success as follows:

> [W]e have developed a global ranking of Web pages called PageRank based on the link structure of the Web that has properties that are useful for search. . . . PageRank is an attempt to see how good an approximation to "importance" can be obtained from just the link structure. [W]e have used PageRank to develop a novel search engine called Google. . . .[19]

Google is novel in that it manages to do a syntactic search for significance by using information about human search to capture the importance of what it finds, without the search algorithm needing to understand the meaning of what is found. As these authors put it:

> The importance of a Web page is an inherently subjective matter, which depends on the readers' interests, knowledge and attitudes. But there is still much that can be said objectively about relative importance of Web pages. . . .

PageRank [is] a method for rating Web pages objectively and mechanically, effectively measuring the human interest and attention devoted to them.[20]

Brin and Page explain:

PageRank relies on the uniquely democratic nature of the Web by using its vast link structure as an indicator of an individual page's value. In essence, Google interprets a link from page A to page B as a vote, by page A, for page B. But, Google looks at considerably more than the sheer volume of votes, or links a page receives; for example, it also analyzes the page that casts the vote. Votes cast by pages that are themselves "important" weigh more heavily and help to make other pages "important." Using these and other factors, Google provides its views on pages' relative importance.[21]

They add:

Of course, important pages mean nothing to you if they don't match your query. So, Google combines PageRank with sophisticated text-matching techniques to find pages that are both *important* and *relevant* to your search. Google goes far beyond the number of times a term appears on a page and examines dozens of aspects of the page's content (and the content of the pages linking to it) to determine if it's a good match for your query.[22]

In the face of the pessimistic conclusion of my informants in 1999 that the Web was growing at such a rate that successful search would soon be impossible, Page, Brin and Winograd showed that their method of search relies precisely on the future growth of the WWW. They note: "An important lesson we have learned . . . is that *size does matter*."[23] That is, where Google is concerned, the more votes as to importance, that is the more hyper-connected websites, the better. Thus, with the arrival of Google, pessimism turned to optimism overnight. Page and Brin conclude: "We are optimistic that our centralized Web search engine architecture will improve in its ability to cover the pertinent text information over time and that there is a bright future for search."[24]

How Far is Distance Learning from Education?

Two

With knowledge doubling every year or so, "expertise" now has a shelf life measured in days; everyone must be both learner and teacher; and the sheer challenge of learning can be managed only through a globe-girdling network that links all minds and all knowledge. I call this new wave of technology hyperlearning. . . . It is not a single device or process, but a universe of new technologies that both possess and enhance intelligence. The hyper in hyperlearning refers not merely to the extraordinary speed and scope of new information technology, but to an unprecedented degree of connectedness of knowledge, experience, media, and brains – both human and nonhuman. . . . We have the technology today to enable virtually anyone who is not severely handicapped to learn anything, at a "grade A" level, anywhere, anytime.

Lewis J. Perelman, **School's Out**, Avon/Education, 1993, pp. 22–3

In 1922 Thomas Edison predicted that "the motion picture is destined to revolutionize our educational system and . . . in a few years it will supplant largely, if not entirely, the use of textbooks". Twenty-three years later, in 1945, William Levenson, the director of the Cleveland public schools' radio station, claimed that "the time may come when a portable radio receiver will be as common in the classroom as the blackboard". Forty years after that, the noted psychologist B. F. Skinner, referring to the first days of his "teaching machines", in the late 1950s and early 1960s, wrote, "I was soon saying that, with the help of teaching machines and programmed instruction students could learn twice as much in the same time and with the same effort as in a standard classroom."[1]

For two decades now computers have been touted as a new technology that will revitalize education. In the 1980s they were proposed as tutors, tutees, and drillmasters but none of those ideas seem to have taken hold.[2] Now the hope is that somehow the power of the World Wide Web will make possible a new approach to education for the twenty-first century in which each student will be able to stay at home and yet be taught by great teachers from all over the world.

Many influential people in the United States believed until recently that the development of the Internet would solve the problems of our current educational system.[3] At the secondary school level, we would no longer have to worry about crammed classes, a deficient infrastructure, or the lowering of standards, and, at the college level, we would be able to leave behind the demographic difficulties posed by too many students, limited access to the most expensive universities, and the need for constant retraining as skill requirements change. If the new technology were put to use in the right way, they maintained, a first-class education would be available to everyone, everywhere, in so far as they mastered the relevant information technology.

Of course, many educators hold the opposite view – namely, that education requires face-to-face interaction between teachers and students. For example, Nancy Dye, President of Oberlin College, is sure that "Learning is a deeply social process that requires time and face-to-face contact. That means professors interacting with students."[4] Likewise, *The New York Times* reports that "the American Federation of Teachers . . . critical of the sterility of distance learning, noted, 'All our experience as educators tells us that teaching and learning in the shared human spaces of a campus are essential to the undergraduate experience.' "[5]

But neither side gives us any reason to accept their pronouncements. In the face of this stand-off with no arguments on either side, we have to take a careful look at education in the light of the new possibilities for distance learning and ask: can distance learning enable students to acquire the skills they need in order to be good citizens skilled in various domains? Or, does learning really require face-to-face engagement, and, if so, why? Just what goes on in classrooms, lecture halls, seminar rooms, and wherever skills are learned?

First, we need to get clear about what skills are and how they are acquired.[6] So, before seeking to evaluate the conflicting claims concerning distance learning, I'll lay out briefly what seem to be the stages in which a student learns by means of instruction, practice, and, finally, apprenticeship, to become an expert in some particular domain and in everyday life and what more is required for one to become a master. The question then becomes: can these stages be implemented and encouraged on the Web?

STAGE 1: NOVICE

Normally, the instruction process begins with the instructor decomposing the task environment into context-free features that the beginner can recognize without the desired skill. The beginner is then given rules for determining actions on the basis of these features, like a computer following a program.

For purposes of illustration, I'll consider three variations: a motor skill, an intellectual skill, and what takes place in the lecture hall. The student automobile driver learns to recognize such domain-independent features as speed (indicated by the speedometer) and is given rules such as shift to second when the speedometer needle points to ten. The novice chess player learns a numerical value for each type of piece regardless of its position, and the rule: "Always exchange if the total value of pieces captured exceeds the value of pieces lost." The player also learns to seek centre control when no advantageous exchanges can be found, and is given a rule defining centre squares and one for calculating extent of control.

In the classroom and lecture hall, the teacher supplies the facts and procedures that need to be learned in order for the student to begin to develop an understanding of some particular domain. The student learns to recognize the features and follow the procedures by drill and practice. As long as students are merely consumers of information, as they are at this stage, they don't need to be in a classroom with each other and a teacher at all. Each can learn at his own terminal, wherever and whenever is convenient. Clearly, in this way the Internet can offer an improved version of the correspondence course, but this can't be what the enthusiasts are shouting about.

In any case, merely following rules will produce poor performance in the real world. A car stalls if one shifts too soon on a hill or when the car is heavily loaded; a chess player who always exchanges to gain points is sure to be the victim of a sacrifice by the opponent who gives up valuable pieces to gain a tactical advantage. Understanding a language or a science is much more than memorizing the elements and the rules relating them. The student needs not only the facts but also an understanding of the context in which that information makes sense.

STAGE 2: ADVANCED BEGINNER

As the novice gains experience actually coping with real situations and begins to develop an understanding of the relevant context, he or she begins to note, or an instructor points out, perspicuous examples of meaningful additional aspects of the situation or domain. After seeing a sufficient number of examples, the student learns to recognize them. Instructional maxims can then refer to these new situational *aspects*, recognized on the basis of experience, as well as to the objectively-defined non-situational *features* recognizable by the novice. Unlike a rule, a maxim requires that one already has some understanding of the domain to which the maxim applies.[7]

The advanced beginner driver uses (situational) engine sounds as well as (non-situational) speed in deciding when to shift. He learns the maxim: shift up when the motor sounds like it's racing and down when it sounds like it's straining. Engine sounds cannot be captured adequately by a list of features. In general, features cannot take the place of a few choice examples in learning the relevant distinctions.

With experience, the chess beginner learns to recognize over-extended positions and how to avoid them. Similarly, she begins to recognize such situational aspects of positions as a weakened king's side or a strong pawn structure, despite the lack of precise and de-situated definitions. The player can then follow maxims such as: attack a weakened king's side.

At school, mere information is contextualized so that the student can begin to develop an understanding of its significance. The instructor takes on the role of a coach who helps the student pick out and recognize the relevant aspects that organize and make sense of the material. Although aspects can be presented to passive students in front of their terminals, it is more effective for the student to attempt to use the maxims that have been given, while the instructor points out aspects of the current situation to the student as the student encounters them. Here the teacher needs to be present with the student in the actual situation of thought or action.

Still, at this stage, as the student follows instructions and is given examples, learning, whether it takes place at a distance or face to face, can be carried on in a detached, analytic frame of mind. But to progress further requires a special kind of involvement.

STAGE 3: COMPETENCE

With more experience, the number of potentially relevant elements and procedures that the learner is able to recognize and follow becomes overwhelming. At this point, since a sense of what is important in any particular situation is missing, performance becomes nerve-racking and exhausting, and the student might well wonder how anybody ever masters the skill.

To cope with this overload and to achieve competence, people learn, through instruction or experience, to devise a plan, or choose a perspective, that then determines which elements of the situation or domain must be treated as important and which ones can be ignored. As students learn to restrict themselves to only a few of the vast number of possibly relevant features and aspects, understanding and decision making becomes easier.

Naturally, to avoid mistakes, the competent performer seeks rules and reasoning procedures to decide which plan or perspective to adopt. But such rules are not as easy to come by as are the rules and maxims given to beginners in manuals and lectures. Indeed, in any skill domain the performer encounters a vast number of situations differing from each other in subtle ways. There are, in fact, more situations than can be named or precisely defined, so no one can prepare for the learner a list of types of possible situations and what to do or look for in each. Students, therefore, must decide for themselves in each situation what plan or perspective to adopt, without being sure that it will turn out to be successful.

Given this uncertainty, coping becomes frightening rather than merely exhausting. Prior to this stage, if the rules don't work, the performer, rather than feeling remorse for his mistakes, can rationalize that he hadn't been given adequate rules. But since, at this stage, the result depends on the perspective adopted by the learner, the learner feels responsible for his or her choice. Often, the choice leads to confusion and failure. But sometimes things work out well, and the competent student then experiences a kind of elation unknown to the beginner.

A competent driver leaving the freeway on an off-ramp curve learns to pay attention to the speed of the car, not whether to shift gears. After taking into account speed, surface condition, criticality of time, etc., he may decide he is going too fast. He then has to decide whether to let up on the accelerator, remove his foot altogether, or step on the brake, and precisely when to perform any of these actions. He is relieved if he gets through the curve without mishap, and shaken if he begins to go into a skid.

The class A chess player, here classed as competent, may decide after studying a position that her opponent has weakened his king's defences so that it makes sense to attack the king. If she chooses to attack, she ignores weaknesses in her own position created by the attack, as well as the loss of pieces not essential to the attack. Pieces defending the enemy king become salient. Since pieces not involved in the attack are being lost, the timing of the attack is critical. If she attacks too soon or too late, her pieces will have been lost in vain and she will almost surely lose the game. Successful attacks induce euphoria, while mistakes are felt in the pit of the stomach.

If we were disembodied beings, pure minds free of our messy emotions, our responses to our successes and failures would lack this seriousness and excitement. Like a computer we would have goals and succeed or fail to achieve them, but, as John Haugeland once said of chess machines that have been programmed to win, they seek their goal, but, when it comes to winning, they don't give a damn. For embodied, emotional beings like us, however, success and failure do matter. So the learner is naturally frightened, elated, disappointed, or discouraged by the results of his or her choice of perspective. And, as the competent student becomes more and more emotionally involved in his task, it becomes increasingly difficult for him to draw back and adopt the detached maxim-following stance of the advanced beginner.

But why let learning be infected with all that emotional stress? Haven't we in the West, since the Stoics, and especially since Descartes, learned to make progress by mastering our emotions and being as detached and objective as possible? Wouldn't rational motivations, objective detachment, and honest evaluation be the best way to acquire expertise?

While it might seem that involvement could only interfere with detached rule-testing, and so would inevitably lead to irrational decisions and inhibit further skill development, in fact, just the opposite seems to be the case. Patricia Benner has studied nurses at each stage of skill acquisition. She finds that, unless the trainee stays emotionally involved and accepts the joy of a job well done, as well as the remorse of mistakes, he or she will not develop further, and will eventually burn out trying to keep track of all the features and aspects, rules and maxims that modern medicine takes account of. In general, resistance to involvement and risk leads to stagnation and ultimately to boredom and regression.[8]

Since students tend to imitate their teachers, teachers can play a crucial role in whether students will withdraw into being disembodied minds or become more and more emotionally involved in the learning situation. If the teacher is detached and computer-like, the students will be too. Conversely, if the teacher shows his involvement in the way he pursues the truth, considers daring hypotheses and interpretations, is open to students' suggestions and objections, and emotionally dwells on the choices that have led him to his conclusions and actions, the students will be more likely to let their own successes and failures matter to them, and rerun the choices that led to these outcomes.

In the classroom and lecture hall the stakes are less dramatic than the risk of having a car accident while driving or of losing an important game of chess. Still, there is the possibility of taking the risk of proposing and defending an idea and finding out whether it fails or flies. If each student is alone in front of his or her computer, there is no place for such risky involvement. Indeed, the correspondence-course model of anonymous information consumers that promoters of distance learning seem to have in mind when they say that the course material will be available to anyone, anywhere, any time, makes such involvement impossible. But, even if we drop the any time, and suppose that the students are all watching the professor at the same time, as with interactive video, and everyone watching hears each student's question, each student is still anonymous and there is still no class before which the student can shine and also risk making a fool of himself. The professor's approving or disapproving response might carry some emotional weight but it would be much less intimidating to offer a comment and get a reaction from the professor if one had never met the professor and was not in her presence. Thus, those who agree with President Dye and the American Federation of Teachers may well be right.

The Net's limitations where embodiment is concerned – the absence of face-to-face learning – may well leave students stuck at competence.

STAGE 4: PROFICIENCY

Only if the detached, information-consuming stance of the novice, advanced beginner, and distance learner is replaced by involvement, is the student set for further advancement. Then, the resulting positive and negative emotional experiences will strengthen successful responses and inhibit unsuccessful ones, and the performer's theory of the skill, as represented by rules and principles, will gradually be replaced by situational discriminations, accompanied by associated responses. Proficiency seems to develop if, and only if, experience is assimilated in this embodied, atheoretical way. Only then do intuitive reactions replace reasoned responses.

As usual, this can be seen most clearly in cases of action. As the performer acquires the ability to discriminate among a variety of situations, each entered into with involvement, plans are evoked and certain aspects stand out as important without the learner standing back and deciding to adopt that perspective. When the perspective is simply obvious, rather than the winner of a complex competition, there is less doubt as to whether what one is trying to accomplish is appropriate.

At this stage, the involved, experienced performer sees each situation from an intuitive perspective, but hasn't yet learned what to do. This is inevitable since there are far fewer ways of seeing situations than there are ways of reacting. The proficient performer simply has not yet had enough experience with the outcomes of the wide variety of possible responses to each of the situations he can now discriminate to react automatically. Thus, the proficient performer, after spontaneously seeing the salient components of the current situation, must still *decide* what to do on the bases of highly salient and less salient, but relevant, components of the situation. And to decide, he must fall back on detached rule- and maxim-following.

The proficient driver, approaching a curve on a rainy day, may *feel in the seat of his pants* that he is going dangerously fast. He must then *decide* whether to apply the brakes or merely to reduce pressure on the accelerator by some specific amount. Valuable time may be lost while making a decision, but the proficient driver is certainly more likely to negotiate the curve safely than the competent driver who spends additional time *considering* the speed, angle of bank, and felt gravitational forces, in order to *decide* whether the car's speed is excessive.

The proficient chess player, who is classed a master, can, if shown a meaningful chess position, intuitively discern almost immediately the salient forces inherent in the situation. She then deliberates to determine what to do. She may know, for example, that she should attack, but she must calculate how best to do so.

A student at this level sees the problem that needs to be solved but has to figure out what the answer is.

STAGE 5: EXPERTISE

The *proficient performer*, immersed in the world of his skilful activity, *sees* what needs to be done, but has to *decide* how to do it. The *expert* not only sees what needs to be achieved; thanks to his vast repertoire of intuitive perspectives, he also sees immediately what to do. The ability to make more subtle and refined discriminations is what distinguishes the expert from the proficient performer. Among many situations, all seen as similar with respect to a plan or a perspective, the expert has learned to distinguish those situations requiring one reaction from those demanding another. That is, with enough experience in a variety of situations, all seen from the same perspective but requiring different tactical decisions, the brain of the expert gradually decomposes this class of perspectives into subclasses, each of which requires a specific response. This allows the immediate intuitive situational response that is characteristic of expertise.

The expert driver not only feels in the seat of his pants when speed is critical and too fast; given the other components of the situation that are experienced as relevant, he knows how to perform the appropriate action without calculating and comparing alternatives. On the off-ramp, his foot simply lifts off the accelerator and applies the appropriate pressure to the brake. What must be done, simply is done.

The chess Grandmaster experiences a compelling intuitive perspective and a sense of the best move. Excellent chess players can play at the rate of 5 to 10 seconds a move and even faster without any serious degradation in performance. At this speed they must depend almost entirely on intuition and hardly at all on analysis and comparison of alternatives. It has been estimated that an expert chess player can distinguish roughly 100,000 types of positions. For expert performance in other domains, the number of intuitive perspectives with associated actions built up on the basis of experience must be comparatively large.

The student, who has mastered the material, immediately sees the solution to the current problem.

Of course, there are special circumstances where detached deliberation *can* prove useful to a human expert, such as when more than one compelling perspective or action intuitively presents itself, or when the situation is recognized as sufficiently novel as to put in doubt an intuitive behaviour learned through only very limited experience. When I now deal with the achievement of mastery, I will identify a role for a different sort of deliberate, effortful behaviour, used even in situations where one is already performing expertly.

What is the role of the teacher at this stage? A student learns by small random variations on what he is doing, and then checking to see whether or not his performance has improved. Of course, it would be better for learning if these small random variations where not random – if they were sensible deviations. If the learner watches someone who is good at doing something, that could limit the learner's random trials to the more promising one's.[9] So observation and imitation of the activity of an expert can replace a random search for better ways to act. In general, this is the advantage of being an apprentice. Its importance is particularly clear in professional schools.

One thing that professional schools must teach is the way the theory the student has learned can be applied in the real world. A way to accomplish this without apprenticeship is for the school to simulate the surroundings that the students are to function in at a later point in their careers. Business schools provide an instructive example. At American schools of business administration two different modes of thought compete. One is to be found in the so-called analytical school where most teaching focuses on theory. This type of school rarely produces capable business people who are intuitive experts. The other tradition is based on case studies, where real-life situations are described to the students and discussed. This produces better results.

To become an expert, however, it is not suffcient to have

worked through a lot of cases. As we have already seen in discussing the move from competence to proficiency, the cases must matter to the learner. Just as flight simulators work only if the trainee feels the stress and risk of the situation and does not just sit back and try to figure out what to do, for the case method to work, the students must become emotionally involved. So, in a business school case study, the student should not be confronted with objective descriptions of situations, but rather be led to identify with the situation of the senior manager and experience his agonized choices and subsequent joys and disappointments. Provided that they draw in the embodied, emotional student, not just his mind, simulations – especially computer simulations – can be useful. The most reliable way to produce involvement, however, is to require that the student work in the relevant skill domain. So we are back at apprenticeship.

Even where the subject matter is purely theoretical, apprenticeship is necessary. Thus, in the sciences, postdoctoral students work in the laboratory of a successful scientist to learn how their disembodied, theoretical understanding can be brought to bear on the real world. By imitating the master, they learn abilities for which there are no rules, such as how long to persist when the work does not seem to be going well, just how much precision should be sought in each different kind of research situation, and so forth. In order to bring theory into relation with practice, this sort of apprenticeship turns out to be essential.

Even in the humanities where there are no agreed-upon theories, the graduate student needs personal guidance. Thus, she normally becomes a teaching assistant where she can interact with a practising teacher. The teacher can't help but exhibit a certain style of approaching texts and problems and of asking questions. For example, he may manifest an aggressive style of never admitting he is wrong or a receptive style of soliciting objections and learning from his mistakes. It is their teacher's style more than anything else that the teaching assistants pick up and imitate, even though they usually don't realize that they are doing so. An inspiring teacher like Wittgenstein left several succeeding generations of students not only imitating his style of questioning but even his gestures of puzzlement and desperation.

For passing-on a style, apprenticeship is the only technique available. However, if what the expert produced were clones of his or her own style, apprenticeship would be stultifying. Taking the notion of apprenticeship seriously, one has to ask how, within this framework, new styles and innovative ability can be developed. The training of musicians provides a clue. If you are training to become a performing musician, you have to work with an already recognized master. The apprentice cannot help but imitate the master, because when you admire someone and spend time with them, their style becomes your style. But then the danger is that the apprentice will become merely a copy of the master, while being a virtuoso performing artist requires developing a style of one's own.

Musicians have learned from experience that those who follow one master are not as creative performers as those who have worked sequentially with several.[10] The apprentice, therefore, needs to leave his first master and work with a master with a different style. In fact, he needs to study with several such masters. Journeymen in medieval times, and performing artists even now, when they are ready to develop a style of their own, travel around and work in various communities of practice. In music, the teachers encourage their students to work with them for a while and then go on to other teachers. Likewise, graduate students usually assist several professors, and young scientists may work in several laboratories.

It is easy for us moderns to misunderstand this need for apprenticeship to several teachers. We tend to think, for example, that the music apprentice needs to go to one master because she is good at fingering, to another because he is good at phrasing, and yet another because she is good at dynamics. That would suggest one could divide a skill into components, which is the wrong way to look at it. Rather, one master has one whole style and another has a wholly different style.[11] Working with several masters destabilizes and confuses the apprentice so that he can no longer simply copy any one master's style and so is forced to begin to develop a style of his own.

STAGE 6: MASTERY

With experience, one can, and generally does, just naturally become what we call an expert. Given enough experience, it is difficult to avoid it. All animals tend to become expert at what survival demands. Paradoxically, it seems that only a human being can be so attached to the deliberative rule-based thinking typical of the first three stages of instructed skill acquisition and so afraid of taking any risk, that vast experience produces only enhanced competence within a skill domain. Also, however, only human beings can become masters.

A very different sort of deliberation from that of a rule-using competent performer or of a deliberating expert characterizes the master. At one level of explanation one can say that the future master consciously decides that expertise isn't good enough. For example, a person might be dedicated to what counts as excellence in her profession and therefore dissatisfied with merely engaging in what is accepted as expert behaviour. In general, unlike the average expert who is satisfied to perform well, to become a master a learner must be strongly motivated to look for opportunities to excel that are invisible to experts and must be willing to accept the risk of temporarily degraded performance while further developing their skill.

How does the developing master find opportunities for improvement that the satisfied expert does not see? To answer, we must first look in some detail at the matter of "perspective" as discussed in presenting stages 3 and 4 of the skill model. Recall that for the advanced beginner an aspect is an experience of a discriminable class of experiences such as the car motor sounding like it's straining, which can be identified and given a name by a teacher, but which cannot be described as a combination of context-free features. For the competent performer, perspective means the deliberative choice of which context-free features and aspects of a situation are important constituents of one's guidelines for behaviour, and which are irrelevant or of lesser significance. For the proficient performer, however, perspective is best thought of as a set of experiences, most of which are unnamable, with some experiences seen as crucial and others as of lesser or no importance.

The future master must be willing and able, in certain situations, to override the perspective that as an expert performer he intuitively experiences. The budding master forsakes the available "appropriate perspective" with its learned accompanying action and deliberately chooses a new one. This new perspective lacks an accompanying action, so that too must be chosen, as it was when the expert was only a proficient performer. This of course risks regression in performance and is generally done during rehearsal or practice sessions. Sometimes a coach, who is himself a master or who has learned to become a masterful coach, will suggest or demonstrate a new way of experiencing a situation, but a new perspective can also be chosen experimentally without coaching by a highly motivated expert. When consciously overriding conventional expertise happens to yield improved performance, the resulting emotionally rewarding experience reinforces the likelihood that, when in a similar situation in the future, the newly established perspective and action will recur without conscious effort, and what might be called "enhanced expertise" results. The strongly motivated aspiring master generally will replay the memory of the rewarding experience many times and do so with the same emotional involvement as accompanied it in the first place. This will help solidify the perspective and behaviour in the learner's repertoire.

A related alternative road to mastery presents itself to experts whose skill demands that they sometimes must respond to novel situations without time for deliberation. Such an expert, if motivated to excel, not only will assess the situation spontaneously and respond immediately, but will experience elation if the assessment and response is successful and dissatisfaction if it seems to him disappointing. But, unlike ordinary satisfied experts, if the developing master is dedicated to his profession and if time permits, he will recall and savour successes. Alternatively, in case of dissatisfaction, there seems to be two possible ways to respond. He may *deliberate* about what should have been done and make a rule to do things a different way if a similar situation arises in the future. He then risks the temporary regression to competence that comes with resisting an intuitive response, but this new way of acting will, hopefully, become intuitive with more experience. Or, rather than analysing what went wrong and making a rule for avoiding the mistake in the future, he may just dwell on the past events, feeling sad about what happened when things went wrong and joy when recalling the times when they went well. Then simple pleasure and pain conditioning will rewire his neurons in a way that will lead him to repeat the successful types of performance and prevent him from acting in the unsatisfactory way in the future. In either case, the new behaviour will become part of the master's ever-growing intuitive repertoire that is activated immediately if a similar situation occurs in the future.

For example, a masterful professional basketball player known for his exceptional ability to pass appropriately the ball to a teammate in a better position to score will have undoubtedly done this many times during practice when honing this skill. He will be dedicated to his chosen sport and will have savoured successes during practice and played them over in his mind after the session. A dedicated craftsperson will try unusual combinations of materials, some of which will be successful and some not, in the process of learning just naturally to use the right materials to create masterpieces. An expert nurse, seeking to develop into a master because of her dedication to caregiving, cannot rely on improvement by trial and error, but she will notice situations where she did the conventional thing and wished after an undesirable outcome that she had done things differently. By dwelling on that situation and imagining with emotional involvement what she might have noticed and then done and how it might have turned out better, she will respond differently and perhaps masterfully in similar situations in the future. Expert professors and lawyers, skilled in a profession that sometimes requires spontaneous responses, have available, if sufficiently dedicated, both the deliberative and the alternative, non-reflective road to mastery that can be used when time permits after the event.

To sum up, when an *expert* learns, she must either create a new perspective in a situation when a learned perspective has failed, or improve the action guided by a particular intuitive perspective when the intuitive action proves inadequate. A *master* will not only continue to do this, but will also, in situations where she is already capable of what is considered adequate expert performance, be open to a new intuitive perspective and accompanying action that will lead to performance that exceeds conventional expertise. Thus, although producing a higher level of skill than the expert, the brain of the master doesn't use any different operating principles. Rather, thanks to exceptional motivation due to their dedication to their chosen profession, the ability to savour and dwell on successes, and a willingness to persevere despite the risk of regression during learning, the master's brain comes to instantiate significantly more available perspectives with accompanying actions than the brain of an expert. Thanks to practice, these perspectives are invoked when they are appropriate, and the master's performance rises to a level of excellence unavailable to the ordinary expert.

Not only do people have to acquire skills by imitating the style of experts in specific domains; they have to acquire the style of their culture in order to gain what Aristotle calls practical wisdom. Children begin to learn to be experts in their culture's practices from the moment they come into the world. In this task, they are apprenticed to their parents from the word go.

Our cultural style is so embodied and pervasive that it is generally invisible to us, so it is helpful to contrast our style with some other cultural style and compare how it is learned. Sociologists point out that mothers in different cultures handle their babies in different ways.[12] For example, American mothers tend to place babies in their cribs by putting them on their stomachs, which encourages the babies to move around. Japanese mothers, contrariwise, put their babies on their backs so they will lie still, lulled by whatever they hear and see. American mothers encourage passionate gesturing and vocalizing, while Japanese mothers are much more soothing and mollifying. In general American mothers situate the baby's body and respond to the baby's actions in such a way as to promote an active and aggressive style of behaviour. Japanese mothers, in contrast, promote a greater passivity and sensitivity to harmony. Thus, what constitutes the American baby as an *American* baby is its style, and what constitutes the Japanese baby as a *Japanese* baby is its quite different style.

The general cultural style determines how the baby encounters himself or herself, other people, and things. Starting with a style, various practices will make sense and become dominant and others will either become subordinate or will be ignored altogether. So, for example, babies never encounter a bare rattle. For an American baby a rattle-thing is encountered as an object to make lots of expressive noise with and to throw on the floor in a wilful way in order to get a parent to pick it up. A Japanese baby may treat a rattle-thing this way more or less by accident, but generally, I suspect, a rattle-thing is encountered as serving a soothing, pacifying function like a Native American rainstick.

Once we see that a style governs how anything can show up *as* anything, we can see that the style of a culture governs not only the babies. The adults in each culture are completely shaped by it too. For example, it should come as no surprise to us, given the sketch of Japanese and American culture already presented, that Japanese adults seek contented, social integration, while American adults are still striving wilfully to satisfy their individual desires. Likewise, the style of enterprises and of political organizations in Japan aims at producing and reinforcing cohesion, loyalty, and consensus, while what is admired by Americans in business and politics is the aggressive energy of a *laissez-faire* system in which everyone strives to express his or her own individuality, and where the state, businesses, or other organizations function to maximize the number of desires that can be satisfied without destructive instability.

Like everyday commonsense understanding, cultural style is too embodied to be captured in a theory, and passed on by talking heads. It is simply passed on silently from body to body, yet it is what makes us human beings and provides the background against which all other learning is possible. Only by being an apprentice to one's parents and teachers can one gain practical wisdom – the general ability to do the appropriate thing, at the appropriate time, in the appropriate way. If we were to leave our bodies behind and live in cyberspace, nurturing children and passing on one's variation of one's cultural style to them would become impossible.

CONCLUSION

At every stage of skill acquisition beyond the first three, involvement and mattering are essential. Like expert systems following rules and procedures, the immortal detached minds envisaged by futurists like Moravec would at best be competent.[13] Only emotional, involved, embodied human beings can become proficient and expert. So, while they are teaching specific skills, teachers must also be incarnating and encouraging involvement. Moreover, learning through apprenticeship requires the bodily presence of masters, and picking up the style of life that we share with others in our culture requires being in the presence of our elders. On this basic level, as Yeats said, "Man can embody the truth, but he cannot know it."[14]

When one looks at education in detail – from hands-on coaching, to manifesting the necessary involvement, to showing how the theory of a domain can be brought to bear on real situations, to developing one's own style, to mastering an activity – one can see how much distance learning leaves out. Indeed, in so far as we want to teach expertise and mastery in particular domains and practical wisdom in life, which we certainly want to do, we finally run up against the most important question a philosopher can ask those who believe in the educational promise of the World Wide Web: can the bodily presence required for acquiring skills in various domains and for acquiring mastery of one's culture be delivered by means of the Internet?

The promise of telepresence holds out hope for a positive answer to this question. If telepresence could enable human beings to be present at a distance in a way that captures all that is essential about bodily presence, then the dream of distance learning at all levels could, in principle, be achieved. But if telepresence cannot deliver a substitution for classroom coaching and lecture-hall presence through which involvement is fostered by committed teachers, as well as the presence to apprentices of masters whose style is manifest on a day-to-day basis so that it can be imitated, distance learning will produce only competence, while expertise, let alone mastery, will remain completely out of reach. Hyper-learning would then turn out to be mere hype. So our question becomes: how much presence can telepresence deliver?

Three

She could see the image of her son, who lived on the other side of the
earth, and he could see her. . . . "What is it, dearest boy?" . . . "I want you
to come and see me." "But I can see you!" she exclaimed. "What more
do you want?" . . . "I see something like you . . . but I do not see you.
I hear something like you through this phone, but I do not hear you."
The imponderable bloom, declared by discredited philosophy to be
the actual essence of intercourse, was ignored by the machine.

E. M. Forster, "The Machine Stops"[1]

Artists see far ahead of their time. Thus, just after the turn of
the last century, E. M. Forster envisioned and deplored an age
in which people would be able to sit in their rooms all their
lives, keeping in touch with the world electronically. Now
we have almost arrived at this stage of our culture. We can
keep up with the latest events in the universe, shop, do
research, communicate with our family, friends, and colleagues, meet new people, play games, and control remote
robots all without leaving our rooms. When we are engaged
in such activities, our bodies seem irrelevant and our minds
seem to be present wherever our interest takes us.[2]

As we have seen, some enthusiasts rejoice that, thanks to
progress in achieving such telepresence, we are on the way to
sloughing off our situated bodies and becoming ubiquitous
and, ultimately, immortal. Others worry that if we stay in our
rooms and only relate to the world and other people through
the Net we will become isolated and depressed, as the

Carnegie-Mellon study mentioned in the Introduction seems to confirm.

A more recent and more extensive study at Stanford University confirmed the isolation but did not take up the question of the loneliness and depression. *The New York Times* reports:

> In contrast to the Carnegie-Mellon study, which focused on psychological and emotional issues, the Stanford survey is an effort to provide a broad demographic picture of Internet use and its potential impact on society. . . . Mr. Nie [the survey director] asserted that the Internet was creating a broad new wave of social isolation in the United States, raising the specter of an atomized world without human contact or emotion.[3]

The Stanford researchers, like the sponsors of the Carnegie-Mellon survey, were surprised by their findings. They lament that no one is trying to look ahead to what, if anything, we will lose if we limit ourselves to disembodied interactions. " 'No one is asking the obvious questions about what kind of world we are going to live in when the Internet becomes ubiquitous', Mr. Nie said."[4] Since that is precisely what we are trying to do in this book, we had better get on with it.

Lovers of the Internet claim that we will soon be able to live our lives through a vast Network that will become more and more dense like a tissue or like an invisible ocean in which we will swim. They see this as a great opportunity. *Wired Magazine* tells us:

> Today's metaphor is the network – a vast expanse of nodes strung together with dark, gaping holes in between. But as the threads inevitably become more tightly drawn, the mesh

will fill out into a fabric, and then – with no voids whatsoever – into an all-pervasive presence, both powerful and unremarkable. . . . In the words of Eric Brewer, a specialist on computer security and parallel computing, it will be "a giant, largely invisible infrastructure that makes your life better".[5]

Given that many people now agree that, as things are going, we will soon live our lives through such a vast, invisible, interconnected infrastructure, we must surely ask: will it, indeed, make our lives better? What would be gained and what, if anything, would be lost if we were to take leave of our situated bodies in exchange for ubiquitous telepresence in cyberspace? We can break up this question into two: how does relating to the world through teletechnology affect our overall sense of reality? And what, if anything, is lost when human beings relate to each other by way of teletechnology? (See Chapter 5.) To answer these questions, we will first have to explore the more general question: what is telepresence and how is it related to our everyday experience of being in the presence of things and people?

In modernity, we tend to ask how can we ever get out of our inner, private, subjective experience so as to be in the presence of the things and people in the external world? While this seems an important question to us now, it was not always taken seriously. The Greeks thought of human beings as empty heads turned towards the world. St Augustine worked hard to convince people that they had an inner life. In his *Confessions* he goes out of his way to comment on the amazing fact that St Ambrose could read to himself. "When he read, his eyes scanned the page and his heart explored the meaning, but his voice was silent and his tongue was still."[6] But the idea that there was an inner world didn't really take hold until early in the seventeenth century when three influences led René Descartes to make the modern distinction between the contents of the mind and the rest of reality.

To begin with, instruments like the telescope and microscope were extending man's perceptual powers, but along with such indirect access came doubts about the reliability of what one seemed to see by means of such prostheses. The church doubted Galileo's report of spots on the sun and, as Ian Hacking tells us, "even into the 1860s there were serious debates as to whether globules seen through a microscope were artifacts of the instrument or genuine elements of living material (they were artifacts)".[7]

At the same time, the sense organs themselves were being understood as transducers bringing information to the brain. Descartes pioneered this research with an account of how the eye responded to light and passed the information on to the brain by means of "the small fibres of the optic nerve".[8] Likewise, Descartes understood that other nerves brought other information to the brain and from there to the mind. He thought that this showed that our access to the world is indirect, that is, that things are never directly present to us, but we experience them by way of representations in our brain and mind.

Descartes then went even further and used reports of people with a phantom limb to call into question our seemingly direct experience of our bodies:

I have been assured by men whose arm or leg has been amputated that it still seemed to them that they occasionally felt pain in the limb they had lost – thus giving me grounds to think that I could not be quite certain that a pain I endured was indeed due to the limb in which I seemed to feel it.[9]

So Descartes concluded that the world and even our own bodies are never directly present to us but that all that we can directly experience is the content of our own minds. And, indeed, when we engage in philosophical reflection, it seems we have to agree with Descartes. It seems to us that we do not have direct access to the external world but only to our private, subjective experiences.

If this were our true condition, then the mediated information concerning distant objects and people transmitted to us over the Internet as telepresence would be as present as anything could get. But, in response to the Cartesian claim that all our experience of the world is indirect, pragmatists such as William James and John Dewey emphasized that the crucial question is whether our relation to the world is that of a disembodied detached spectator or an involved embodied agent. On their analysis, what gives us our sense of being in direct touch with reality is that we can control events in the world and get perceptual feedback concerning what we have done.

But even this sort of control and feedback is not sufficient to give the controller a sense of direct contact with reality. As long as we are controlling a robot with delayed feedback, such as Ken Goldberg's Telegarden arm[10] or the Mars Sojourner, what we see on the screen will seem to be mediated by our long-distance equipment, and therefore not truly tele-*present*.

There comes a point in interactive robot control, however, where we are able to cope skillfully with things and people in real time. Then, as in laparoscopic-surgery, for example, the doctor feels himself present at the robot site, the way blind people feel themselves present at the end of their cane. But even though interactive control and feedback may give us a sense of being directly in touch with the objects we manipulate, it may still leave us with a vague sense that we are not in touch with reality. Something about the distance still undermines our sense of direct presence.

One might think that what is missing from our experience as we sit safely at home remotely controlling our car, for example, is a constant readiness for risky surprises. To avoid extremely risky situations is precisely why remotely-controlled planet-exploring vehicles and tools for handling radioactive substances were developed in the first place; but, in the everyday world, our bodies are always in potentially risky situations. So, when we are in the real world, not just as minds but as embodied vulnerable human beings, we must constantly be ready for dangerous surprises. Perhaps, when this sense of vulnerability is absent, our whole experience is sensed as unreal, even if, involved in a sort of super-Imax interactive display, we are swaying back and forth as our car careens around dangerous-looking curves. But aren't believers in the triumph of technology such as the Extropians right on this point? Couldn't we develop a technologically-controlled world so tame that being on our guard all the time was no longer necessary? And wouldn't it still seem real?

Maurice Merleau-Ponty has attempted to answer this question, and refute Descartes, by describing just what gives us our sense of the world being directly present to us. He holds that there is a basic need we can never banish as long as we have bodies. It is the need to get what Merleau-Ponty calls an optimal grip on the world. When grasping something, we tend to grab it in such a way as to get the best grip on it. Merleau-Ponty points out that, in general, when we are looking at something, we tend, without thinking about it, to find the best distance for taking in both the thing as a whole and its different parts. Merleau-Ponty says:

> For each object, as for each picture in an art gallery, there is
> an optimum distance from which it requires to be seen: . . . at
> a shorter or greater distance we have a perception blurred
> through excess or deficiency. We therefore tend towards the
> maximum of visibility, and seek a better focus as with a
> microscope.[11]

According to Merleau-Ponty, it is the body that seeks this
optimum:

> My body is geared into the world when my perception
> presents me with a spectacle as varied and as clearly
> articulated as possible, and when my motor intentions, as
> they unfold, receive the responses they expect from the world.
> This maximum sharpness of perception and action points
> clearly to a perceptual ground, a basis of my life, a general
> setting in which my body can co-exist with the world.[12]

So, perception is motivated by the indeterminacy of experience and our perceptual skills serve to make determinable
objects sufficiently determinate for us to get an optimal grip
on them. Moreover, we wouldn't want to evolve beyond the
tendency of our bodies to move so as to get a grip on the
world since this tendency is what leads us to organize our
experience into the experience of stable objects in the first
place. Without our constant sense of the uncertainty and
instability of our world and our constant moving to overcome
it, we would have no stable world at all.[13]

Not only is each of us an active body coping with things,
but, as embodied, we each experience a constant readiness to
cope with things in general that goes beyond our readiness
to cope with any specific thing. Merleau-Ponty calls this
embodied readiness our Urdoxa[14] or "primordial belief" in

the reality of the world. It is what gives us our sense of the direct presence of things. So, for there to be a sense of presence in telepresence, one would not only have to be able to get a grip on things at a distance; one would need to have a sense of the context as soliciting a constant readiness to get a grip on whatever comes along.

This sense of being embedded in a world with which we are set to cope is easiest to see if we contrast our experience of the direct presence of other people with telepresence such as teleconferencing. Researchers developing devices for providing telepresence hope to achieve a greater and greater sense of actually being in the presence of distant people and events by introducing high-resolution television and surround sound, and by adding touch and smell channels. Scientists agree that "full telepresence requires a transparent display system, high resolution image and wide field of view, a multiplicity of feedback channels (visual as well as aural and tactile information, and even environmental data such as moisture level and air temperature), and a consistency of information between these".[15] They assume that the more such multi-channel, real-time, interactive coupling teletechnology gives us, the more we will have a sense of the full presence of distant objects and people.

But even such a multi-channel approach may not be sufficient. Two roboticists at Berkeley, John Canny and Eric Paulos, criticize the attempt to break down human–human interaction into a set of context-independent communication channels such as video, audio, haptics, etc. They point out that two human beings conversing face to face depend on a subtle combination of eye movements, head motion, gesture, and posture and so interact in a much richer way than most roboticists realize.[16] Their studies suggest that a holistic sense of embodied interaction may well be crucial to everyday human encounters, and that this *intercorporiality*, as Merleau-Ponty calls it, cannot be captured by adding together 3D images, stereo sound, remote robot control, and so forth.

Just what is missing can best be seen if we return to the question of distance learning. We ended the last chapter by asking whether the presence of the teacher required for full-fledged learning could be captured by telepresence. We are now in a position to suggest an answer to this question. But, rather than looking at the six stages of skill acquisition from the point of view of the learner, we will look at learning from the point of view of the teacher and ask, what, if anything, does the teacher lose in attempting to teach skills at a distance?

If the teacher is only recording videotape, then there is no telepresence at all, and a great deal is surely lost. For example, if risk is important in the learning process, then when the teacher and the class are present together *both* assume a risk that is not there when they are not interacting – the student risks being called on to demonstrate his knowledge of the subject of the lecture, and the teacher risks being asked a question he cannot answer. If this is the case, then it may mean that distance teaching not only may produce poorer learning opportunities, but it may produce poorer teaching.

It's true that we think of teachers teaching students, but it is also the case that in an interactive classroom environment the students teach the teacher. The teacher learns that certain examples do or do not work, that some material has to be presented differently from others, that he was simply wrong about some fact or theory, or even that there was a better way of looking at the whole question. It's been said that a "good university" is one that has teachers and learners, but that a "great university" has only learners. If so, passive distance education, by removing the risk in learning and teaching, deprives students and teachers of what is most important, namely, learning how to learn.

The challenging case is live, interactive, video distance learning, although this is not the use of the Web that administrators find cost-effective and therefore attractive. Still, it is the sort of technology that could produce telepresence if anything can. David Blair has given a great deal of thought to his experience both in the presence of students in the classroom and in interactive teleteaching. Here are some of his observations.

> In the first place I am often aware of a lot of things going on in the class in addition to a student actually asking a question or commenting. Sometimes when a student asks a question I can see, peripherally, other students nodding their heads in agreement with the question. This would indicate that the student's question is important to the rest of the class so I will take more care in answering it fully. At the other end of the attention spectrum, I can often see, again, peripherally, when students are bored or sleeping or chatting amongst themselves. This means I may have to pick up the pace of the lecture and try to regain their attention. In teaching students at a distance, I can't control where the camera points and what it zooms in on, the way I control what attracts my experienced attention when the class is in front of me.
>
> Second, as I lecture, I'm drawn to the point of view that is most comfortable or informative for me – a point of view that may be different from lecture to lecture and even may change during a lecture. Perhaps this is similar to Merleau-Ponty's notion of "maximum grip". To find this point of view requires

that I be able to move around during the lecture sometimes approaching the students closely, sometimes moving away.

Finally, much of my sense of the immediate presence of the students in a class comes from my ability to make eye contact with them. My experience with the CU-CMe ("see-you-seeme") technology on computers is that you cannot make eye contact over a visual channel, no matter how good the transmission is. To look into another person's eyes, I would have to look straight into the camera but then I would not be able to see the eyes of the other person since, to do that, I would have to turn from the camera to the student's image on the screen. You can look into the camera or look at the screen, but you can't do both.[17]

What is lost, then, in teleteaching and in telepresence in general is the possibility of my controlling my body's movement so as to get a better grip on the world.

What is also lost, even in interactive video, is a sense of the context. In teaching, the context is the mood in the room. In general, mood governs how people make sense of what they are experiencing. Our body is what enables us to be attuned to the current mood. Ask yourself, if you were a telespectator at a party, would you be able to share the mood? Whereas, if you are *present* at a party, it is hard to resist sharing the elation or depression of the occasion.[18] Likewise, there is always some shared mood in the classroom and it determines what matters – what is experienced as exciting or boring, salient or marginal, relevant or irrelevant. The right mood keeps students involved by giving them a sense of what is important.

Like a good teacher, Blair is sensitive to the mood in his classroom. He writes:

As I became more experienced lecturing, I began to have a

sense of the class as not just a collection of students but as a whole – as a single entity. I feel that the class as a whole is attentive, or responsive, or not responsive, or friendly, or skeptical, etc. This feeling is not just the sum of certain students who appear this way, but is a kind of general feeling. I can get this feeling without a sense of any individual students exemplifying these characteristics. I don't think that any telecommunications device could enable me to get that feeling when viewing the audience at a distance.[19]

One can, perhaps, get a sense of the importance of the sort of subtle interactions that Blair so aptly describes by considering the fact that people pay around $60 a seat to go to a play, even though they can see a movie for a fifth as much. This obviously has something to do with being in the presence of the actors. Presumably, the actors, like good lecturers, are, at every moment, subtly and largely unconsciously adjusting to the responses of the audience and thereby controlling and intensifying the mood in the theatre. Thus, the co-presence of audience and performer provides the audience with the possibility of direct interaction with the performer, and it seems clear that it is this communication going on between the performers and the audience that brings the show to life. Also the spectator in the theatre can choose whom to zoom in on, while in a film that choice is made by the director. Thus, the theatre spectator is actively involved in what happens in front of him, and this contributes to his sense of being present in the same world as the actors.

This way of looking at the importance of bodily presence raises a new question. Films and CDs are different from plays and concerts but each, in its own way, seems just as gripping as its embodied counterpart. Clearly, some stage actors can learn to act in movies, and some live performers can succeed as studio musicians able to produce an intense effect without any feedback from an audience. It should be possible, then, for a lecturer to use the feedback from the cameras and microphones that show remote students, to involve those students in the lecture, without his needing to manage the mood in the remote rooms. This possibility can't be excluded a priori. We will just have to wait and see if distance education breeds a new brand of teleteachers – teacher-movie-actors who are as effective as the current teacher-live-performers.

Still, if we follow the movie/play comparison to the end, the idea that the teleteacher could equal the powerful effect of a skilled teacher who is present in the same room with her students seems unlikely. Without the sense of the mood in the room as well as the shared risk, the involvement of the students with a movie-actor teacher will almost surely be less intense than that of students and teachers reacting to each other's presence. So, it seems that, given the skill model I proposed in the previous chapter, in the domain of education at least, each technological advance that makes teaching more economical and more flexible, by making the teacher and student less immediately present to each other, makes the teaching less effective. One would expect to see a decline in involvement and effectiveness, from tutorial teaching to classroom teaching, to large lecture halls, to interactive video, to asynchronous Net-based courses.

Given this trade off of economy and efficacy, it looks like we might well end up with a two-tiered educational system where those who can afford it will pay five times as much as the distance learning students pay, in order to be in the presence of their professors. This would amount to an elitism similar to the English elitism of Oxford and Cambridge *vis-à-vis* the other universities that don't have tutorials – the very elitism that the democratic levelling produced by distance learning is supposed to eliminate.

The inferiority of distance learning at the college level seems clear, but what about the vocational and postgraduate teaching which is thought to be the forte of the Internet? One study of the advantages of continuous education on the Internet typifies the jargon and the misplaced optimism characteristic of the field.

> Distributed education encompasses distance education but reaches further to imagine a global disaggregation of instructional resources into modular components of excellence which can be reassembled by any organization in the "business" of certifying quality-assured learning accomplishment (certificates and degrees). The result should be a conveniently and affordably accessible, enriched educational environment that integrates the networked delivery of learningware and asynchronous and synchronous conversations within learning communities of student apprentices, their expert mentors, and their educational and career advisors.[20]

Such claims completely miss the point of mentoring and apprenticeship. As we have already seen, the role of the master is to pass on to the apprentice the ability to apply the theory of some domain in the real world. But, one might well ask, why not just record the master at work and transmit his image to his teleapprentices? For example, why not just put a camera on the head of a doctor teaching interns on his rounds and wire him with a microphone so that the teleinterns can see and hear just what the doctor and the interns who are present see and hear?

What, if anything, would the teleinterns miss? The answer again is immersion in the situation. A camera fixed to the doctor's forehead would, indeed, look wherever he focused his attention, so the teleinterns might well see even better than those actually present in the hospital what the doctor was currently seeing. But the problem is that it is the doctor's responsiveness to the whole situation that determines which details he pays attention to and zooms in on. The camera on the doctor's head would, thus, show distant students exactly which feature of the patient's condition the doctor was looking at, but not the background that led that feature to stand out for the doctor so that he zoomed in on it. The teleintern would surely learn something from a televised image of what the doctor pays attention to, but he or she would always remain a prisoner of the doctor's attention setting, just as in a telelecture the professor is a prisoner of the camera operator and the sound engineer in the distant lecture hall. Yet the ability to zoom in on what is significant is one of the most important skills the intern diagnostician has to learn.

So why not also have a camera and microphone that record and transmit the whole ambient hospital scene? The distance-intern could then watch, on a split screen, both the background and what the doctor focuses attention on, and so learn to notice those features of the overall scene that solicit the doctor's attention.

Here, as in the lecture-hall case, the devil is in the phenomenological details. For the doctor who is actually involved in the situation, it's not as if he had two views – one, a wide-angle view of the uninterpreted situation, and the other, a close-up of the details he is focused on. In becoming a diagnostic master, the doctor has learned to see an already-interpreted situation where certain features and aspects spontaneously stand out as meaningful, just as, as one becomes familiar with a strange city, it ceases to look like a jumble of buildings and streets and develops what Merleau-Ponty calls a familiar physiognomy. The intern is trying, among other things, precisely to acquire the doctor's physiognomic perceptual understanding.

So why, if the intern sees the correlation between the uninterpreted scene on half the screen and the relevant features on the other, couldn't he acquire the doctor's physiognomic understanding? Precisely because the technology deprives the learner of bodily involvement in a risky real environment where he has to interpret the scene himself and learn from his mistakes. Merleau-Ponty would argue that, if one does not have the experience of zooming in on the details that, on the basis of previous experience, come to elicit one's attention, and then discovering the hard way when one is right and when one is mistaken as to the relevant details, one will not find that the scene becomes more and more full of meaning. Thus, the distance-apprentice will not learn to respond to the overall scene by being drawn to zoom in on what is significant. But this is precisely what the intern must learn if he is to become an expert diagnostician.

In the real learning situation, where the patient, the doctor, and the interns are directly present, the apprentice doctors can shift their attention to new details they take to be significant and then find out whether they were right or missed something important. If they are thus involved, then, with every success and failure, the overall organization of their background changes, so that in future encounters a different aspect will stand out as significant. There is thus a constantly enriched interaction between the details and the overall significance of the situation. Merleau-Ponty calls this kind of feedback between one's actions and the perceptual world, the intentional arc.[21] And he points out that it functions only if the perceiver is using his body as an "I can", that is, in this case, if he controls where he looks.

So, to learn to see what the doctor sees, the teleintern must be able to control the direction each camera points and how much each camera zooms in or out. After all, simply by having a great deal of passive experience, by watching football games on TV, for example, one can become competent at following the ball and even predicting and interpreting the plays. So one might well think that adding control of what the camera focused on would enable the tele-student to acquire an expert feel for any skill domain. In such an ideal distance-learning setup, would anything required for learning still be left out?

As we saw in Chapter 2, the learner becomes an expert by reacting to specific situations, and taking to heart the results. On the basis of sufficient such experience, the brain of the beginner gradually comes to connect perception and action so that, in a situation similar to one that has already been experienced, the agent immediately makes a response similar to the response that worked the last time the learner was in that type of situation. But this requires that the learning situations in which one acquires a skill be sufficiently similar to actual situations so that the responses one learns in training carry over to the real world.

So, any form of telelearning, whether interactive or not, must face a final challenge. Can telepresence reproduce the sense of being in the situation so that what is learned transfers to the real world? Experienced teachers and phenomenologists agree that the answer is "no". To see in a stark and extreme form the sort of embodied presence any attempt to transmit full presence cannot capture, it helps to take an example from a physical sport like football.

Barry Lamb, Safeties Coach for the Brigham Young University Football Team and a former All-American linebacker and defensive end at Santa Barbara Community College (1973–74), reports the following:

Our players can learn a great deal by watching films, but only to a point. It's hard to say exactly what it is that you can't learn by watching film, but a good player learns to sense the overall situation and to do things instinctively that just don't make sense if you're only looking at what you can see on film. Most game film, of course, is not taken from a player's perspective. But even if you could correct for that, the depth of field is never the same on film as it is in real life.[22] That means that you can't really learn to see the playing field in the right way, or get a feel for the tempo of the game. In addition, there is more to learning how to see a play develop than just having your head or eyes pointed in the right direction. Our players need to learn how to use their peripheral vision to get a feel for what is going on around them, and what your peripheral vision tells you makes you see what is going on in front of you differently.[23] Moreover, the emotions of the game change how a player sees the field, and those aren't things that one can get a feel for from the film.

Another way to see how the film is too sterile to teach everything our players need to learn is by noticing that opposing players aren't threatening on film in the same way that they are in real life. The fact that there are eleven players in front of you who want to hurt you really makes you see and understand things differently.

In sum, learning to do the right thing, a thing that

sometimes doesn't make sense, is something that can only happen when a person experiences a present situation over and over again, whether in practice or in real life.[24]

All this suggests that distance-learners looking at a surround screen and hearing stereo sound would be able to develop a degree of competence. Thus, an intern could become competent at recognizing and, perhaps, even anticipating many of the symptoms the doctor has pointed out, just as an avid TV viewer can learn to recognize and anticipate many of the plays on the footfall field. Furthermore, if the learner could view the scene transmitted by cameras placed exactly where the actual embodied learner would normally be placed, he might even be able to become proficient. But such distance-learners would still lack the experience that comes from responding directly to the risky and perceptually rich situations that the world presents. Without an experience of their embodied successes and failures in actual situations, such learners would not be able to acquire the ability of an expert or a master who responds immediately to present situations in a masterful way. So we must conclude that expertise cannot be acquired in disembodied cyberspace. Distance-learning enthusiasts notwithstanding, apprenticeship can only take place in the shared situations of the home, the hospital, the playing field, the laboratory, and the production sites of crafts. Distance-apprenticeship is an oxymoron.

Once we see that there is a way of being directly present to things and people that is denied by Descartes and all of modern philosophy, we see that there may well be basic limitations on telepresence that go far beyond the problems of distance teaching. Where the presence of people rather than objects is concerned, we sense a crucial difference between those we have access to through our distance senses of hearing, sight, etc. and the full-bodied presence that is literally within arm's reach. This full-bodied presence is more than the feeling that I am present at the site of a robot arm I am controlling from a distance through real-time interaction. Nor is it just a question of giving robots surface sensors so that, through them as prostheses, we can touch other people at a distance. Even the most gentle person–robot interaction would never be a caress, nor could one successfully use a delicately controlled and touch-sensitive robot arm to give one's kid a hug. Whatever hugs do for people, I'm quite sure telehugs won't do it. And any act of intimacy mediated by any sort of robot prosthesis would surely be equally grotesque, if not obscene. Even if our teletechnology goes beyond the imagination of E. M. Forster so that eventually we can use remote-controlled robotic arms and hands to touch other people, I doubt that people could get a sense of how much to trust each other even if they could stare into each other's eyes on their respective screens, while, at the same time, using their robot arms to shake each other's robotic hands.

Perhaps, one day, we will stop missing this kind of bodily contact, and touching another person will be considered rude or disgusting. E. M. Forster envisions such a future in his story:

When Vashti swerved away from the sunbeams with a cry [the flight attendant] behaved barbarically – she put out her hand to steady her. "How dare you!" exclaimed the passenger, "you forget yourself!" The woman was confused, and apologized for not having let her fall. People never touched one another. The custom had become obsolete, owing to the Machine.[25]

For the time being, however, investment bankers know that in order to get two CEOs to trust one another enough to merge their companies, it is not sufficient that they have many tele-conferences. They must live together for several days inter-acting in a shared environment, and it is quite likely that they will finally make their deal over dinner.[26]

What is the connection between such trust and embodied presence? Perhaps our sense of trust must draw on the sense of security and well-being each of us presumably experienced as babies in our caretaker's arms.[27] Our sense of reality, then, would not be just the readiness for flight of a hunted animal; it could also be the feeling of joy and security of being pro-tected. If so, even the most sophisticated forms of telepresence may well seem remote and even obscene if not in some way connected with our sense of the warm, encircling, nearness of an actual human body.

Of course, there are many kinds of trust, and the trust that we have that our mail carrier will deliver our mail does not require looking her in the eye or shaking her hand. The kind of trust that requires such body contact is our trust that some-one will act sympathetically to our interests even when so doing might go against his or her own.[28]

So, it seems that to trust someone you have to make your-self vulnerable to him or her and they have to be vulnerable to you. Part of trust is based on the experience that the other does not take advantage of one's vulnerability. You have to be in the same room with someone who could physically hurt or publicly humiliate you and observe that they do not do so, in order to trust them and make yourself vulnerable to them in other ways.

There is no doubt that telepresence can provide some sense of trust, but it seems to be a much-attenuated sense. Perhaps in the future world of the Internet we will none the less come to prefer telepresence to total isolation, like Harlow's monkeys who, lacking a real mother, shun the wire "mother" and cling desperately to the terry-cloth one – never knowing the comfort and security of a real mother's arms.[29]

Not that we automatically trust anyone who hugs us. Far from it. Just as for Merleau-Ponty it is only on the background of our embodied faith in the presence and reality of the perceptual world that we can doubt the reality of any specific perceptual object, so we seem to have a background predisposition to trust those who touch us tenderly, and it is only on the basis of this *Urtrust* that we can be mistrustful in any specific case. If that background trust were missing, as it would necessarily be in cyberspace, we might tend to be suspicious of the trustworthiness of every social interaction and withhold our trust until we could justify it. Such a scepticism would complicate if not poison all human interaction.

CONCLUSION

We have now seen that our sense of the reality of things and people and our ability to interact effectively with them depend on the way our body works silently in the background. Its ability to get a grip on things provides our sense of the reality of what we are doing and are ready to do; this, in turn, gives us a sense both of our power and of our vulnerability to the risky reality of the physical world. Furthermore, the body's ability to zero in on what is significant, and then preserve that understanding in our background awareness, enables us to perceive more and more refined situations and respond more and more skillfully; its sensitivity to mood opens up our shared social situation and makes people and things matter to us; and its tendency to respond positively to direct engagement with other bodies; underlies our sense of trust and so sustains our interpersonal world. All this our body does so effortlessly, pervasively, and successfully that it is hardly noticed. That is why it is so easy to think that, thanks to telepresence, we could get along without it, and why it would, in fact, be impossible to do so.

Four

> Oh God said to Abraham, "Kill me a son" . . .
> Well Abe says, "Where do you want this killin' done"
> God says, "Out on Highway 61".
> Well Mack the Finger said to Louie the King
> I got forty red white and blue shoe strings
> And a thousand telephones that don't ring
> Do you know where I can get rid of these things
> And Louie the King said let me think for a minute son.
> And he said yes I think it can be easily done
> Just take everything down to Highway 61.
>
>   Now the rovin' gambler he was very bored
> He was tryin' to create a next world war
> He found a promoter who nearly fell off the floor
> He said I never engaged in this kind of thing before
> But yes I think it can be very easily done
> We'll just put some bleachers out in the sun
> And have it on Highway 61.
>
> Bob Dylan, "Highway 61 Revisited"

In a section of *A Literary Review* entitled "The Present Age",[1] written in 1846, Kierkegaard warns that his age is character-ized by a disinterested reflection and curiosity that level all differences of status and value. In his terms, this detached reflection levels all qualitative distinctions. Everything is equal in that nothing matters enough that one would be willing to die for it. Nietzsche gave this modern condition a name; he called it nihilism.

Kierkegaard blames this levelling on what he calls the

Public. He says that "For levelling properly to come about a phantom must first be provided, its spirit, a monstrous abstraction, an all-encompassing something that is a nothing, a mirage – this phantom is the *public*."[2] But the real villain behind the Public, Kierkegaard claims, is the Press. He warned that "Europe will come to a standstill at the Press and remain at a standstill as a reminder that the human race has invented something which will eventually overpower it",[3] and he adds: "Even if my life had no other significance, I am satisfied with having discovered the absolutely demoralizing existence of the daily press."[4]

But why blame levelling on the public rather than on democracy, technology, or loss of respect for tradition, to name a few candidates? And why this monomaniacal demonizing of the press? Kierkegaard says in his journals that "Actually it is the Press, more specifically the daily newspaper . . . which make[s] Christianity impossible."[5] This is an amazing claim. Clearly, Kierkegaard saw the press as a unique cultural/religious threat, but it will take a little while to explain why.

It is no accident that, writing in 1846, Kierkegaard chose to attack the public and the press. To understand why he did so, we have to begin a century earlier. In *The Structural Transformation of the Public Sphere*[6] Jürgen Habermas locates the beginning of what he calls the *public sphere* in the middle of the eighteenth century. He explains that at that time the press and coffeehouses became the locus of a new form of political discussion. This new sphere of discourse was radically different from the ancient polis or republic; the modern public sphere understood itself as being outside political power. This extrapolitical status was not just defined negatively, as a lack of political power, but seen positively. Just because public opinion is not an exercise of political power, it is protected from any partisan spirit. Enlightenment intellectuals saw the public sphere as a space in which the rational, disinterested reflection that should guide government and human life could be institutionalized and refined. Such disengaged discussion came to be seen as an essential feature of a free society. As the press extended public debate to a wider and wider readership of ordinary citizens, Burke exalted that, "in a free country, every man thinks he has a concern in all public matters".[7]

Over the next century, thanks to the expansion of the daily press, the public sphere became increasingly democratized until this democratization had a surprising result which, according to Habermas, "altered [the] social preconditions of 'public opinion' around the middle of the [nineteenth] century."[8] "[As] the Public was expanded . . . by the proliferation of the Press . . . the reign of public opinion appeared as the reign of the many and mediocre."[9] Many people, including J. S. Mill and Alexis de Tocqueville, feared "the tyranny of public opinion"[10] and Mill felt called upon to protect "nonconformists from the grip of the Public itself".[11] According to Habermas, Tocqueville insisted that "education and powerful citizens were supposed to form an *elite public* whose critical debate determined public opinion".[12]

"The Present Age" shows just how original Kierkegaard was. While Tocqueville and Mill claimed that the masses needed elite philosophical leadership, and while Habermas agrees with them that what happened around 1850 with the democratization of the public sphere by the daily press is an unfortunate decline into conformism from which the public sphere must be rescued, Kierkegaard sees the public sphere itself as a new and dangerous cultural phenomenon in which the nihilism produced by the press brings out something that was deeply wrong with the Enlightenment idea of detached reflection from the start. Thus, while Habermas is concerned to recapture the moral and political virtues of the public sphere, Kierkegaard warns that there is no way to salvage the public sphere since, unlike concrete and committed groups, it was from the start the source of levelling.

This levelling was produced in several ways. First, the new massive distribution of desituated information was making every sort of information immediately available to anyone, thereby producing a desituated, detached spectator. Thus, the new power of the press to disseminate information to everyone in a nation led its readers to transcend their local, personal involvement and overcome their reticence about what didn't directly concern them. As Burke had noted with joy, the press encouraged everyone to develop an opinion about everything. This is seen by Habermas as a triumph of democratization, but Kierkegaard saw that the public sphere was destined to become a detached world in which everyone had an opinion about and commented on all public matters without needing any first-hand experience and without having or wanting any responsibility.

The press and its decadent descendant, the talk show, are bad enough, but this demoralizing effect was not Kierkegaard's main concern. For Kierkegaard, the deeper danger is just what Habermas applauds about the public sphere, namely, as Kierkegaard puts it, "the public . . . eats up all individuality's relativity and concreteness".[13] The public sphere thus promotes ubiquitous commentators who deliberately detach themselves from the local practices out of which specific issues grow and in terms of which these issues must be resolved through some sort of committed action. As Kierkegaard puts it:

The public is not a people, a generation, one's era, not a community, an association, nor these particular persons, for all these are only what they are by virtue of what is concrete. *Not a single one of those who belong to the public has an essential engagement in anything.*[14]

What seems a virtue to detached Enlightenment reason, therefore looks like a disastrous drawback to Kierkegaard. Even the most conscientious commentators don't have to have first-hand experience nor take a concrete stand. Rather, as Kierkegaard complains, they justify their views by citing principles. Since the conclusions such abstract reasoning reaches are not grounded in the local practices, its proposals would presumably not enlist the commitment of the people involved, and consequently would not work even if enacted as laws.

More basically still, that the public sphere lies outside of political power meant, for Kierkegaard, that one could hold an opinion on anything without having to act on it. He notes with disapproval that "[the public's] artifice, its good sense, its virtuosity consists in letting matters reach a verdict and a decision without ever acting".[15] This opens up the possibility of endless reflection. For, if there is no need for decision and action, one can look at all things from all sides and always find some new perspective. The accumulation of information thus postpones decision indefinitely since, as one finds out more, it is always possible that one's picture of the world, and, therefore, of what one should do, will have to be revised. Kierkegaard saw that, when everything is up for endless critical commentary, action can always be postponed. "[R]eflection is able at any moment to put things in a new light and allow one some measure of escape."[16] One need never act.

All that a reflective age like ours produces is more and more knowledge. As Kierkegaard puts it, "One can say in general of a passionless but reflective age, compared to a passionate one, that it *gains in extensity what it loses in intensity*."[17] He adds: "we all know what path to take and what paths can be taken, but no one will take them."[18] No one stands behind the views the public holds, so no one is willing to act on them. Kierkegaard wrote in his journal: "here . . . are the two most dreadful calamities which really are the principle powers of impersonality – the Press and anonymity."[19] Therefore, the motto Kierkegaard suggested for the press was: "Here men are demoralized in the shortest possible time on the largest possible scale, at the cheapest possible price."[20]

In "The Present Age" Kierkegaard succinctly sums up his view of the relation of the press, the public sphere, and the levelling going on in his time. The desituated and anonymous press and the lack of passion or commitment in his reflective age combine to produce the public, the agent of the nihilistic levelling:

> The abstraction of the press (for a newspaper, a journal, is no political concretion and only an individual in an abstract sense), combined with the passionlessness and reflectiveness of the age, gives birth to that abstraction's phantom, the public, which is the real leveller.[21]

Kierkegaard would surely have seen in the Internet, with its Websites full of anonymous information from all over the world and its interest groups that anyone in the world can join without qualifications and where one can discuss any topic endlessly without consequences, the hi-tech synthesis of the worst features of the newspaper and the coffeehouse.[22] Indeed, thanks to the Internet, Burke's dream has been realized. In news groups, anyone, anywhere, any time, can have an opinion on anything. All are only too eager to respond to the equally deracinated opinions of other anonymous amateurs who post their views from nowhere. Such commentators do not take a stand on the issues they speak about. Indeed, the very ubiquity of the Net tends to make any such local stand seem irrelevant.

The most perfect realization of Burke's vision of the public sphere is the blog. In a blog anyone can express his or her opinion about anything without needing any experience and without accepting any responsibility. But since putting one's ideas into practice and so taking risks and learning from one's failures and successes are required for acquiring expertise, most bloggers have no wisdom to contribute.

The Enlightenment hope is that the few bloggers who are engaged in the concrete activities they write about will be recognized and be widely read, but the flood of blogs, the fact that those involved in committed action are generally too busy to write commentaries, and the fact that the readers who are supposed to do the job of recognizing the enlightening blogs by clicking on them are not themselves experienced and wise, makes the contribution of blogs to serious public debate unlikely. Blogging is more interactive than the press and talk shows, and so resembles a return to the coffeehouse kibitzing of those outside of power that Habermas admires as democracy at work, and Kierkegaard detests as a diversion that substitutes for risky action.

What Kierkegaard envisaged as a consequence of the press's indiscriminate and uncommitted coverage is now fully realized on the World Wide Web. Thanks to hyperlinks, meaningful differences have, indeed, been levelled. Relevance and significance have disappeared. And this is an important part of the attraction of the Web. Nothing is too trivial to be included. Nothing is so important that it demands a special place. In his religious writing Kierkegaard criticized the implicit nihilism in the idea that God is equally concerned with the salvation of a sinner and the fall of a sparrow,[23] that "for God there is nothing significant and nothing insignificant".[24] He said such a thought would lead one "to the verge of despair".[25] On the Web, the attraction and the danger are that everyone can take this godlike point of view. One can view a coffee pot in Cambridge, or the latest super-nova, study the Kyoto Protocol, find out what fellowships are available to a person with one's profile, or direct a robot to plant and water a seed in Austria, not to mention plough through thousands of ads, all with equal ease and equal lack of any sense of what is important. The highly significant and the absolutely trivial are laid out together on the information highway in just the way Abraham's sacrifice of Isaac, red, white and blue shoe strings, a thousand telephones that don't ring, and the next world war are laid out on Dylan's nihilistic "Highway 61".

Kierkegaard even foresaw that the ultimate activity the Internet would encourage would be speculation on how big it is, how much bigger it will get, and what, if anything, all this means for our culture. This sort of discussion is, of course, in danger of becoming part of the very cloud of anonymous speculation Kierkegaard abhorred. Ever sensitive to his own position as a speaker, Kierkegaard concluded his analysis of the dangers of the present age and his dark predictions of what was ahead for Europe with the ironic remark that: "And since in this age, in which so little is actually done, such an extraordinary amount is done in the way of prophecies, apocalypses, hints, and insights into the future, there is probably nothing else for it but to join in."[26]

The only alternative Kierkegaard saw to the public's levelling and paralysing reflection was for one to plunge into some kind of activity – any activity – as long as one threw oneself into it with passionate commitment. In "The Present Age" he exhorts his contemporaries to make such a leap:

> There is as little action and decision these days as shallow-water paddlers having a dangerous desire to swim. But just as the adult being tossed about delightedly by the waves calls to the younger person: "Come on, just jump right in" – so the decision so to speak lies within existence (but in the individual, naturally) and shouts to the younger person not yet exhausted by an excess of reflection . . .: "Come on, jump boldly in." Even if it were a reckless leap, so long as it is decisive – if you have it in you to be a man, then the danger and life's stern judgment upon your recklessness will help you become one.[27]

Such a light-hearted leap out of the shallow, levelled present age into the deeper water is typified for Kierkegaard by people who leap into what he calls the *aesthetic sphere of existence*. Each sphere of existence, as we shall see, represents a way of trying to get out of the levelling of the present age by making some way of life absolute.[28] In the aesthetic sphere, people make enjoyment of all possibilities the centre of their lives.

Such an aesthetic response is characteristic of the Net-surfer for whom information gathering has become a way of life. Such a surfer is curious about everything and ready to spend every free moment visiting the latest hot spots on the Web. He or she enjoys the sheer range of possibilities. For such a person, just visiting as many sites as possible and keeping up on the cool ones is an end in itself. The qualitative distinction that staves off levelling for the aesthete is the distinction between those sites that are *interesting* and those that are *boring*, and, thanks to the Net, something interesting is always only a click away. Life consists in fighting off boredom by being a spectator at everything interesting in the universe and in communicating with everyone else so inclined. Such a life produces what we would now call a postmodern self – a self that has no defining content or continuity but is constantly taking on new roles.

But we have still to explain what makes this use of the Web so attractive. Why is there a thrill in being able to be up on everything no matter how trivial? What motivates a passionate commitment to curiosity? Kierkegaard thought that people were addicted to the press, and we can now add the Web, because the anonymous spectator *takes no risks*. The person in the aesthetic sphere keeps open all possibilities and has no fixed identity that could be threatened by disappointment, humiliation, or loss.

Life on the Web is ideally suited to such a mode of existence. On the Internet, commitments are at best virtual commitments. Sherry Turkle has described how the Net is changing the background practices that determine what kinds of selves we can be. In *Life on the Screen*, she details "the ability of the Internet to change popular understandings of identity". On the Internet, she tells us, "we are encouraged to think of ourselves as fluid, emergent, decentralized, multiplicious, flexible, and ever in process".[29] Thus "the Internet has become a significant social laboratory for experimenting with the constructions and reconstructions of self that characterize postmodern life".[30]

Chat rooms lend themselves to the possibility of playing at being many selves, none of whom is recognized as who one truly is, and this possibility is not just theoretical but actually introduces new social practices. Turkle tells us that:

> The rethinking of human . . . identity is not taking place just among philosophers but "on the ground", through a philosophy in everyday life that is in some measure both proved and carried by the computer presence.[31]

She notes with approval that the Net encourages what she calls "experimentation" because what one does on the Net has no consequences.[32] For that very reason, the Net frees people to develop new and exciting selves. The person living in the aesthetic sphere of existence would surely agree, but according to Kierkegaard: "As a result of knowing and being everything possible, one is in contradiction with oneself."[33] When he is speaking from the point of view of the next, higher, sphere of existence, Kierkegaard tells us that the self requires not "variableness and brilliancy" but "firmness, balance, and steadiness".[34]

We would therefore expect the aesthetic sphere to reveal that it was ultimately unliveable, and, indeed, Kierkegaard holds that, if one leaps into the aesthetic sphere with total commitment expecting it to give one's life meaning, it is bound to break down. Without some way of telling the significant from the insignificant and the relevant from the irrelevant, everything becomes equally interesting and equally boring and one finds oneself back in the indifference of the present age. Writing from the perspective of an aesthete experiencing the despair that signals the breakdown of the aesthetic sphere, Kierkegaard laments: "My reflection on life altogether lacks meaning. I take it some evil spirit has put a pair of spectacles on my nose, one glass of which magnifies to an enormous degree, while the other reduces to the same degree."[35]

This inability to distinguish the trivial from the important eventually stops being thrilling and leads to the very boredom the aesthete Net-surfer dedicates his life to avoiding. So, if one throws oneself into it fully, one eventually sees that the aesthetic way of life just doesn't work to overcome levelling. Kierkegaard calls such a realization, despair. Thus, Kierkegaard concludes: "every aesthetic view of life is despair, and everyone who lives aesthetically is in despair whether he knows it or not. But when one knows it a higher form of existence is an imperative requirement."[36]

That higher form of existence Kierkegaard calls the ethical sphere. In it, one has a stable identity and one engages in involved action. Information is not played with, but is sought and used for serious purposes. As long as information gathering is not an end in itself, whatever reliable information there is on the Web can be a valuable resource serving serious concerns. Such concerns require that people have life plans and take up serious tasks. They then have goals that determine what needs to be done and what information is relevant for doing it.

In so far as the Internet can reveal and support the making and maintaining of commitments for action, it supports life in the ethical sphere. But Kierkegaard would probably hold that the huge number of interest groups on the Net committed to various causes, and the ease of joining and leaving such groups, would eventually bring about the breakdown of the ethical sphere. The multiplicity of causes and the ease of making and breaking commitments, which should have supported action, will eventually lead either to paralysis or an arbitrary choice as to which commitments to take seriously.

To avoid arbitrary choice, one might, like Judge William, Kierkegaard's pseudonymous author of the description of the ethical sphere in *Either/Or*, turn to facts about one's life to limit one's commitments. Thus, Judge William says that his range of possible relevant commitments is constrained by his abilities, and his social roles as judge and husband. Or, to take a more contemporary example, one could choose which interest groups to join on the basis of certain facts about one's life-situation. After all, there are not merely interest groups devoted to everything from bottle caps to culture stars like Kierkegaard,[37] there are interest groups, for example, for the parents of children with rare and incurable diseases. So the ethical Net-enthusiast might argue that, to avoid levelling, all one need do is to choose to devote one's life to something that matters based on some accidental condition in one's life.

But the goal of the person in the ethical sphere, as Kierkegaard defines it, is to be morally mature, and Kant held that moral maturity consists in the ability to act lucidly *and freely*. To live ethically, then, one cannot base the meaning of one's life on what accidental facts impose their importance. Thus Judge William is proud of the fact that, as an autonomous agent, he is free to give whatever meaning he chooses to his talents and his roles and all other facts about himself. Thus, he claims that, in the end, his freedom to give his life meaning is not constrained by his talents and social duties, unless he chooses to make them important.

Judge William sees that the choice as to which facts about his life are important is based on a more fundamental choice of what is worthy and not worthy, what is good and what is evil, and that choice is up to him. As Judge William puts it:

> The good is for the fact that I will it, and apart from my willing, it has no existence. This is the expression for freedom. . . . By

this the distinctive notes of good and evil are by no means
belittled or disparaged as merely subjective distinctions. On
the contrary, the absolute validity of these distinctions is
affirmed.[38]

But Kierkegaard would respond that, if everything were up
for choice, including the standards on the basis of which one
chooses, there would be no reason for choosing one set of
standards rather than another.[39] Besides, if one were totally
free, choosing the guidelines for one's life would never
make any serious difference, since one could always choose to
rescind one's previous choice. A commitment does not get
a grip on me if I am always free to revoke it.[40] Indeed, commitments that are freely chosen can and should be revised
from minute to minute as new information comes along. The
ethical thus breaks down in despair because, either I am stuck
with whatever just happens to be imposed on me as important in my life (for example, some life-threatening disease)
and so I'm not free, or else the pure power of the freedom
to make and unmake commitments undermines itself. As
Kierkegaard puts the latter point:

If the despairing self is active . . . it is constantly relating to
itself only experimentally, no matter what it undertakes,
however great, however amazing and with whatever
perseverance. It recognizes no power over itself; therefore in
the final instance it lacks seriousness. . . . The self can, at any
moment, start quite arbitrarily all over again.[41]

Thus the *choice* of qualitative distinctions that was supposed
to support serious action undermines it, and one ends up in
what Kierkegaard calls the despair of the ethical. One can take
over some accidental fact about one's life and make it one's

own only by freely *deciding* that it is crucially important, but then one can equally freely decide it is not, so in the ethical sphere all meaningful differences are levelled by one's making one's freedom absolute.

According to Kierkegaard, one can only stop the levelling of commitments by being *given* an individual identity that opens up an individual world. Fortunately, the ethical view of commitments as freely entered into and always open to being revoked does not seem to hold for those commitments that are most important to us. These special commitments are experienced as gripping our whole being. Political and religious movements can grip us in this way, as can love relationships and, for certain people, such "vocations" as science or art. When we respond to such a summons with what Kierkegaard calls infinite passion – that is, when we respond by accepting an *unconditional commitment* – this commitment determines what will be the significant issue for us for the rest of our life. Such an unconditional commitment thus blocks levelling by establishing qualitative distinctions between what is important and trivial, relevant and irrelevant, serious and playful in my life. Living by such an irrevocable commitment puts one in what Kierkegaard called the *Christian/ religious sphere of existence.*[42]

But, of course, such a commitment makes one vulnerable. One's cause may fail. One's lover may leave. The detached reflection of the present age, the hyperflexibility of the aesthetic sphere, and the unbounded freedom of the ethical sphere are all ways of avoiding one's vulnerability, but it turns out, Kierkegaard claims, that, for that very reason, they level all qualitative distinctions, and end in the despair of meaninglessness. Only an unconditional commitment and the strong identity it produces can give an individual a world organized by that individual's unique qualitative distinctions, but such a world is always in danger of destruction.

This leads to the perplexing question: what role if any can the Internet play in encouraging and supporting unconditional commitments? A first suggestion might be that the movement from stage to stage would be facilitated by living experimentally on the Web, just as flight simulators help one learn to fly. One would be solicited to throw oneself into Net surfing until one found that boring, then into freely choosing which interest group was important until that choice revealed its absurdity, and so finally one would be driven to let oneself be drawn into a risky unconditional commitment as the only way out of despair. Indeed, at any stage, from looking for all sorts of interesting Websites as one surfs the Net, to striking up a conversation in a chat room, to joining an interest group to deal with an important problem in one's life, one might just find oneself being drawn into a lifetime commitment. No doubt this might happen – people who meet in chat rooms may fall in love – but it is highly unlikely.

Kierkegaard would surely argue that, while the Internet, like the public sphere and the press, does not *prohibit* unconditional commitments, in the end, it *undermines* them. Like a simulator, the Net manages to capture everything but the risk.[43] Our imaginations can be drawn in, as they are in playing games and watching movies, and no doubt, if we are sufficiently involved to feel we are taking risks, such simulations can help us acquire skills, but in so far as games work by temporarily capturing our imaginations in limited domains, they cannot simulate serious commitments in the real world. Imagined commitments hold us only when our imaginations are captivated by the simulations before our ears and eyes. And that is what computer games and the Net offer us. But the risks are only imaginary and have no long-term consequences.[44] The temptation is to live in a world of stimulating images and simulated commitments and thus to lead a simulated life. As Kierkegaard says of the present age, "it transforms the task itself into an unreal feat of artifice, and reality into a theatre".[45]

The test as to whether one had acquired an unconditional commitment would come only if one had the passion and courage to transfer what one had learned on the Net to the real world. Then one would confront what Kierkegaard calls "the danger and life's stern judgment". But precisely the attraction of the Net, like that of the press in Kierkegaard's time, is that it inhibits that final plunge. Indeed, anyone using the Net who was led to risk his or her real identity in the real world would have to act against the grain of what attracted him or her to the Net in the first place.

So it looks like Kierkegaard may be right. The press and the Internet are the ultimate enemy of unconditional commitment, but only the unconditional commitment of what Kierkegaard calls the religious sphere of existence can save us from the nihilistic levelling launched by the Enlightenment, promoted by the press and the public sphere, and perfected in the World Wide Web.

Virtual Embodiment: Myths of Meaning in
Second Life

Five

The most philosophically fascinating phenomenon so far made possible by the Internet is a virtual world called *Second Life* – a three-dimensional virtual environment one can log on to from one's home computer. There are now over eleven million people signed up as "residents" of that world. Of these, in December 2007, 518,947 spent over one hour a day on-line, and, as of that date, users had spent a total of 25,646,287 hours in *Second Life* since its launch.[1]

Residents visit art galleries, shop for virtual goods, go to concerts, have cybersex, worship, attend classes, have conversations, buy and sell real estate, and so forth. The Vatican has taken on the task of saving souls there[2] and Sweden has opened a virtual Embassy to sign up residents to become tourists in real Sweden.

Philip Rosedale, founder and CEO of Linden Lab, the creator of *Second Life*, writes in Chapter One of *second life: the official guide*:

> You are the one who determines what *Second Life* means to you. Do you enjoy meeting people online, talking to them and doing things together in real time? Welcome to *Second Life*. Do you enjoy creating stuff and making it come alive? Welcome to *Second Life*. Do you enjoy running a business and making money – real money? Welcome to *Second Life*.[3]

These remarks call for a brief overview of the uses of *Second Life* in order to situate and focus on the philosophically interesting one.

(1) *Business ventures*

One can make real money in *Second Life* by starting one's own virtual business. Entrepreneurs hope to earn Linden dollars (the currency of *Second Life*) so as to convert their Linden dollars into US dollars. (The exchange rate fluctuates around 260 Linden dollars to one US dollar.) Established enterprises such as Coca-Cola, Sears, Wells Fargo, IBM, BP, and Toyota are open for business in *Second Life*, and other businesses are rushing to follow. There is some question, however, as to whether this trend will continue. In a sober article in *Wired*, Frank Rose explains why he is not impressed:

> [M]ore than 85 percent of the avatars [figures representing residents in the virtual world] created have been abandoned. Linden's in-world traffic tally, which factors in both the number of visitors and time spent, shows that the big draws . . . are free money and kinky sex. On a random day in June, the most popular location was Money Island (where Linden dollars . . . are given away gratis), with a score of 136,000. Sexy Beach, one of several regions that offer virtual sex shops, dancing, and no-strings hookups, came in at 133,000. The Sears store on IBM's Innovation Island had a traffic score of 281; Coke's Virtual Thirst pavilion, a mere 27.[4]

In any case, the business use of *Second Life* turns it into an extension of everyday life where the issue is making a profit, not whether the commodities exchanged are virtual or real. The crossover from the virtual to the real may be surprising, but it isn't what is philosophically interesting about a virtual environment.

(2) *Playing* Second Life *as a game*

One could stay inside the world of *Second Life* and enjoy it as a role-playing game, but *Second Life* isn't itself a game. The mainstream games provide a structure and narrative that define the actions necessary for advancement. In *Second Life* as in the real world, however, there is no overall goal and so there is no way of ranking the success of those involved. *The official guide* tells us: "It's completely up to you to say whether your second life is a success, and how you came to that decision. And it's completely up to you as to when the experience begins and ends"(300). Thus the world of *Second Life* and games like *World of Warcraft* are worlds apart.

(3) *Building a world*

Building, maintaining, and expanding a virtual world is no doubt a daily challenge at Linden Labs. This fascinating type of work was presciently described from the point of view of a master programmer named Hiro Protagonist in Neal Stephenson's futuristic bestseller, *Snow Crash*.[5] In his account of a future dystopia, Stephenson introduced the idea of a virtual world he called a metaverse, and the term is still used in *Second Life*'s self-description. But, obviously, building a virtual world is a real-world occupation; not the job of those who dwell in the metaverse that Linden's programmers create and maintain.

Yet everything in *Second Life* is a program, and so *Second Life* provides the tools and tutorials that enable residents to contribute to the content of the virtual world. Indeed, the users create almost all of the content in *Second Life*. Rosedale writes to the readers of the official guide: "If *Second Life* is a world at all, it's because you've created it. . . . You add millions of objects to *Second Life* – in the form of cars, clothes, castles, and every other kind of thing you can imagine" (iv).

But, according to the official guide, the vast majority of those enjoying *Second Life* do not regard the programming required to produce things in the virtual world as an end in itself; rather they take it as a necessary access to the virtual goods and services the programming provides. Consequently, a whole industry has grown up in which programmers produce and sell on eBay the programs that provide residents of *Second Life* with the virtual things they desire. Rosedale notes: "You spend close to $5 million . . . every month . . . on the things that other users have created and added to the world"(iv).

(4) *Recovering a sense of enchantment*

Edward Castronova, an influential exponent of the virtues of what he calls synthetic worlds, thinks that the fans of virtual worlds are seeking and finding re-enchanted worlds. Castronova's term "re-enchantment" harks back to Max Weber, who argued in 1917 that modern science had led to a disenchantment of the world. This disenchantment meant that no otherworldly forces are evoked in understanding our world and predicting what will happen in it. Fairies, witches, demons, angels, and the occult are nothing but superstitions and literary imaginings. Science can, in principle, master all things. Weber argued that this transformation of the world into a causal mechanism has left many inhabitants of the modern world with an unaccountable feeling of loss. Those disappointed by the disenchanted nature revealed by natural science but disinclined to return to traditional religion are forced to seek re-enchantment elsewhere.

Castronova maintains that the gods and goblins that are programmed by the residents in *Second Life* and by the game developers of alternative worlds such *World of Warcraft* give the user a new sense of wonder in the face of the supernatural. He notes:

> In the long run we are not able to live without myths, . . . and when we see the ongoing migrations of people into lands where magic has finally been credibly (if crudely) rediscovered, we learn how hungry for myth we have become.

And he suggests:

> [P]erhaps synthetic worlds have begun to offer a new mythology. Perhaps this mythology will eventually be successful, credible, even sublime, so that we will find ourselves in an Age of Wonder.[6]

Unfortunately, this claim misses completely what has been lost. To experience the enchantment of the world means to experience being in the grip of mysterious powers that have authority over you. That sort of power is expressed in traditional myths but it is necessarily lacking in the programmed gods and goblins we wilfully invent and can completely command and understand. Only if powers we have not invented and do not control were to well up and dominate us could we recover a sense of wonder and the sacred.[7]

(5) *Artistic creation*

The official guide tells us:

> Virtual hedonism is fun, but do not let it blind you to other possible SL activities. For many residents, *Second Life*

> primarily represents a great opportunity to develop their
> talents as creators and artists (13).

Residents design clothing and buildings, write poems and books, compose music, and make paintings and movies.

Of all the activities in *Second Life*, these activities are the most impressive but also the least indebted to the unreality of the virtual world. The very same creative activities requiring the same artistic talents, skills, and hard work could have been engaged in the real world. Except for the clothing, sculptors, and buildings, the resulting artistic productions in either world are real, not virtual.[8] The creative activity adds grace and beauty to the world of *Second Life* and sometimes evokes reactions that verge on wonder. They make *Second Life* worth visiting, but these achievements don't give rise to new philosophical questions or insights.

(6) *Finding new friends*

There are many lonely isolated souls whose geographical location or physical condition makes it hard for them to find kindred souls to relate to. These people enjoy the way *Second Life* allows them to meet and converse with people all over the world. In this case *Second Life* functions as a three-dimensional chat room in which the setting and the avatars [the residents' virtual bodies] make the conversational experience more realistic and engaging. However, there is a tension between the goal of the lonely people who are geographically isolated and who would presumably prefer to know the appearance of the real people they are interacting with, and the goal of those whose physical condition is a barrier to conversation and who therefore enjoy the possibility of acting as if they were in a masquerade, presenting themselves through avatars that resemble not how they really look but how they would like to appear.

This tension adds a dimension of uncertainty that can be tantalizing or exasperating depending on one's goal, but it does not pose a philosophical problem. The originators of *Second Life* can leave it to the participants to work out how realistically they present themselves. If residents desire honest interactions they can use the voice mode of communication rather than profit from the anonymity of instant messaging, the usual mode of communication in *Second Life*. In any case, finding new friends can be an important positive function of a metaverse.

[7] *Living in an alternative world*

Second Life also offers the possibility of spending one's time in a virtual world that may be more exciting than the real one. That raises the question of how much of one's life should be spent enjoying an admittedly unreal world. Such a question is so new that, so far as I know, only a few philosophers have pondered it; but *Star Trek* has.[9] In *Star Trek: Generations* Picard tries to enlist the aid of Kirk, who has long ago retired to a holodeck-like virtual world.[10] Picard finds Kirk jumping challenging chasms on a handsome horse. He reminds Kirk that, although the horse and scenery are magnificent and the chasms daunting, the whole set-up is virtual so there is no real risk. Thus, no courage is required and no thrill and satisfaction can be experienced. After thinking it over, Kirk sees Picard's point and returns with him to the risky real world.

However, the use of virtual worlds to express oneself in new ways and experiment with other possible lives could be of great interest to philosophers. Indeed, a few philosophers have sought to describe better possible lives than those offered by our current world. Martin Heidegger has tried to capture what life at its best was, and might again be, by studying the enchanted world of the Homeric Greeks and their relation to their gods, while Friedrich Nietzsche imagined a world after the death of God in which higher human beings whom he calls "free spirits" would engage in constant creativity, enjoying transformation for its own sake. Now, for the first time, philosophers have access to a real virtual world so to speak in which they can take up residence and investigate other styles of life that once were possible or could become possible. One could then compare the satisfactions and disappointments of such different lives.

THE EXISTENTIALIST CRITIQUE OF *SECOND LIFE*

The ever-increasing number of people who spend an average of four hours a day in Second Life don't seem to be tempted to return more than is necessary to their everyday lives. Clearly, the Picard story misses something attractive to most people about virtual worlds.

The drawbacks of our world are obvious. The boundedness and fallibility of individual and group perspective, physical and mental suffering, and the vulnerability of one's world to collapse – all of which we might call our essential finitude – are ineliminable. Blaise Pascal, the first existential thinker, writing in the middle of the 17th century, spells out what he calls our wretchedness:

> Nothing is so insufferable to man as to be completely at rest
> . . . He then feels his nothingness, his forlornness, his
> insufficiency, his dependence, his weakness, his emptiness.
> There will immediately arise from the depth of his heart
> weariness, gloom, sadness, fretfulness, vexation, despair.[11]

One could try to confront the world we are thrown into, to face up to our situation, and to struggle to live in a way that accepts and incorporates our vulnerability without despair, but Pascal goes on to point out that "[a]s men are not able to fight against death, misery, ignorance, they have taken it into their heads, in order to be happy, not to think of them at all".[12] Pascal calls this escapist approach *diversion* and gives as examples indulging in billiards, tennis, gambling, and hunting.[13]

Now, however, the Internet and the virtual worlds it makes possible offer us diversions on a much grander scale. Indeed, thanks to virtual worlds like *Second Life*, we can forget our finitude and immerse ourselves in a rich, safe metaverse. Thus we now face a clear choice between a captivating life of diversion, which existential philosophers like Pascal consider empty and inauthentic, and the authentic life they favour in which one is called to face up to the vulnerability of all one cares about and yet, at the same time, find something meaningful to which to dedicate one's life.

At the limit the question becomes: how much misery should one confront? When would it be preferable and ethically permissible to be under the illusion that one was free of finitude? *Star Trek* has raised this question too. In contrast with Picard's rescue of Kirk in *Generations*, consider the 1964 *Star Trek* episode "The Cage". There Spock has to decide whether or not to "rescue" Captain Pike, whose body has been terribly deformed in an accident, and who is living in a dream world thanks to the Talosians who are masters of illusion. Spock decides to let Pike remain in his virtual world, young and handsome, dallying with the beautiful image of a fellow deformed crash victim.

In this extreme case, illusion may well be a wise choice.

Diversion only looks obviously wrong if one holds that facing the truth is our highest duty, or, more specifically, believes like Pascal that we are all called by God (or, as Martin Heidegger would say, our ontological conscience) to take on the hard work, risk, and sacrifice required in answering our calling. After all, we do admire those, like Franklin Roosevelt, Itzak Perlman, or Stephen Hawking, who, instead of identifying with an invulnerable avatar and diverting themselves by enjoying virtual successes, have struggled with their disabilities in order to respond to the call of something that matters crucially to them and gives their life meaning.

AN ALTERNATIVE WAY OF LIFE ENCOURAGED IN *SECOND LIFE:* EXPLORING NEW WORLDS THROUGH SAFE EXPERIMENTATION

But there may well be more admirable uses of *Second Life* than diversion. One can see *Second Life* as offering a quest rather than a distraction. As a new medium for exploring other ways of life, virtual worlds may enable people to learn through safe experimentation which sort of life works best for them.

Thus, many of the residents of *Second Life* are attracted by the way an alternative world promises to enable them to discover and satisfy their deepest desires. One can, for example, devote one's life to the endless production and consumption of commodities – anything that one can buy and enjoy without any risk or any special skill. *The official guide* says: "Shopping, of course, is one of the most popular activities in *Second Life*" (300). Indeed, in *Second Life* people can use the Linden dollars they are given to acquire all the commodities they desire. There is on offer virtual designer clothes, real estate, cars, houses, furniture, hi-tech gadgets, sex toys, art objects, islands, and so forth – anything that has a price.

But the creators of *Second Life* seem to suspect that collecting commodities as a way of life is not enough to make life worth living. *The official guide* goes out of its way to assure us that "*Second Life* has become more than just a machine to support sellers and buyers" (207), and in an interview Rosedale explains:

> [T]here's initially a desire to just have everything that you've ever wanted: to be very beautiful, to be very sociable, and to be very engaged in a kind of fast-forward version of consumption as we know it in the real world.
>
> But that's the first couple of months. And then after that you've almost reached a Zen-like state where you can say, "Well, I've done everything, but what more is there?" Then you start to ask questions like, "Well, maybe I just want to build a temple on a hill and meditate." [This would presumably have to be real meditation, not virtual meditation.] Or, "I want to contribute to a community. . . ."[14]

So why do people give up on fast-forward consumerism? Perhaps because they have learned that a lot of what is most rewarding in life can't be commodified. Residents of *Second Life* seem to have discovered something like this for themselves. Artemis Cain, one of the residents of *Second Life* quoted in *the official guide*, asks: "Do you want to spend money on all sorts of gadgets, or do you want to create, explore, and try all sorts of different things?" (19).

The official guide takes it to be an advantage of the virtual world that in it breakdowns are generally a lot less serious than in ours. When your second life is not going well, you can simply abandon the troublesome situation – your fickle friend, your lost love, even your avatar body and your identity. What you do has fewer consequences than it would have in the real world, thus you are free to make commitments with fewer risks.

This ease of getting out of sticky situations enables experimentation. In *The official guide* we are told:

> *Second Life* is often held up as the perfect place to get your fantasy on – and yes, there's no other place like it for becoming something you aren't, or even for working out just what it is you want to be. In a sense, it's the epitome of the "walled garden", a place where reality dare not intrude (301).

The attraction of such noncommittal involvements becomes more understandable if one thinks of *Second Life* as a masquerade. In a masquerade, people are disguised and are allowed to do normally forbidden things without adverse consequences for their everyday lives. *Second Life* is much richer and more engaging than a masquerade, but the attraction and essential superficiality of the risk-free carnevalesque relation to reality is the same. In *Second Life* if one breaks up with a lover, one does not have to see the suffering of an actual person or worry about the shock of running into him or her again. After the failure of a virtual marriage one does not have to go through a real divorce. When one's virtual business fails in the virtual world one doesn't have to face bankruptcy. In general, one doesn't have to clean up the mess one leaves. You can always just walk away.

But *the official guide* hastens to point out that in *Second Life* just walking away from a situation you don't like is not the only possible response:

> The right thing to do, of course, is not to leave the world, but simply find something that you *do* like. There's no shortage of choices – shopping, visiting art galleries, skydiving [but with

no risk and so no thrill), bowling [but virtual bowling would presumably require only hand/eye coordination, and so give no full-bodied sense of accomplishment], and attending live shows and concerts are just some of the options available (14, my reservations in brackets).

In general, we are told that

one of the biggest differences between real and virtual life . . . is the amount of control you have over your existence. Virtual life offers you total control of everything – you even choose when to enter the world and when to leave, an ability that's sadly lacking in real life. You are truly the master of your destiny (196).

Although it seems exaggerated to claim that in the virtual world one's essential vulnerability can be eliminated altogether, at least you can enter the virtual world without prior attachments or responsibilities and, when you exit, leave behind whatever attachments and responsibilities you formed there. If, however, you become involved in what you are doing, even in the virtual world you are no longer in total control. Failure in your virtual emotional, professional, or practical life is still always possible. Yet, in a virtual world as in the life the ancient Stoics advocated, the kind of life you lead, including how much vulnerability you accept, is up to you.

But, as usual there is a trade-off. Although risk-free experimentation with ways of life and forms of involvement is more exciting and revealing than consumerism, one could argue that it does not give one serious satisfaction. What, then, might be missing?

TWO RISKY WAYS OF LIFE DISCOURAGED IN *SECOND LIFE:*
BOLD EXPERIMENTATION AND UNCONDITIONAL
COMMITMENT

Nietzsche might sound like he is praising the virtues of *Second Life* in claiming that one can and should constantly be reinventing one's self. He boasts:

> We ourselves keep growing, keep changing, we shed our old bark, we shed our skins every spring, we keep becoming younger, fuller of future, taller, stronger.[15]

But Nietzsche famously also says:

> [B]elieve me: the secret for harvesting from existence the greatest fruitfulness and the greatest enjoyment is to *live dangerously*! Build your cities on the slopes of Vesuvius! Send your ships into uncharted seas!![16]

Nietzsche is saying that a way of living that is exciting and rewarding must be more risky than cautiously trying out new ways of life in a protected garden. In the real world experimentation has serious consequences. It takes courage to try new things since one must be ready and willing to learn from surprising and upsetting consequences. Thus, what makes role-playing easy and risk-free, limits the sort of openness to surprising and dangerous new situations that could lead to real discovery.

A Nietzschean life of daring undertakings and willingness to risk failure is possible in *Second Life*, but *Second Life* does not encourage such risks. Indeed, Nietzsche's call for bold experimentation flies in the face of the supposed advantages of a virtual world. Nietzsche would claim that, while the safe experimentation of *Second Life* is easy and can give you superficial satisfactions as in a synthetic Mardi Gras, only a bold experiment with the real possibility of having to deal with the consequences of failure could help you discover what is really possible and worthwhile for you.

In the end, however, Nietzsche advocates a life of the sort that *Second Life* offers. It is a life free of the dark side of finitude – a life that is

> self-sufficient, rich, liberal with happiness and good will; . . .
> [that] does not permit the petty weeds of grief and chagrin to
> come up at all.[17]

But Søren Kierkegaard would argue that a life free of the possibility of grief and humiliation is also a life free of bliss and glory. According to Kierkegaard, the true opposite of a Nietzschean life of bold but existentially safe constant transformation is a life of immutable commitment. In such a life, you hear a calling just for you and live in terms of it the rest of your life, giving up what you *want* to do for what you are *called* to do. In Chapter 4, I call this making an unconditional commitment. Kierkegaard presents a Christian argument that only a life of unconditional commitment with the work and risk that it requires can save one from despair. A hard-earned skill for which one has made a life of sacrifices, or a love that defines what matters in one's world, or an enterprise to which one has dedicated oneself, give life maximal meaning. But at the same time such commitments make one vulnerable to accidents, humiliation, and grief. Thus, in answering a calling one must be ready to risk everything for what defines who one is. One is, however, then aligned with and blessed by an authority greater than any merely human authority, be it a god, history, a tradition, a lover, or something else that our practices show us is worth our total devotion.

Nat Goldhaber, an early exponent of the virtues of

disembodied existence,[18] points out that we don't have to believe the official guide as to what is possible in Second Life. He then describes a case in which, thanks to the lack of seriousness in Second Life that makes noncommittal experimentation attractive, a person is drawn into a serious unconditional commitment:

> Initially, people may experiment furiously to settle on a "way of being" in *Second Life* that satisfies and stimulates them; a way of life which they feel better represents who they are than the body and position they occupy in the physical and social world. Once they have found this place, this new way of being, they can become deeply invested in it. So deeply that their investment in their body and circumstance in the physical world pales by comparison. With such a commitment, even absent the physical body, there is great risk. There is room for rejection by their peers. There is the possibility of embarrassment. There is the possibility of financial collapse.[19]

Goldhaber clearly sees that finitude, in this case vulnerability, is a necessary aspect of our most meaningful experiences and relationships. And he rightly points out that such vulnerability is possible in Second Life. But granted that finding one's vocation is the most valuable gift one could hope for from real life or from Second Life, it is also important to realize that being drawn into an unconditional commitment is not the normal result of entering the world of easy experimentation. Choosing to live in Second Life is not neutral. According to the official guide, "What is best about Second Life . . . is [that] practically all the restraints and limitations of real life are absent" (194). Second Life does, indeed, enable one to try out a whole spectrum of lives, but it makes activities ranging from consumerism to risk-free experimentation so attractive that it lures one to pursue a life that minimizes vulnerability and maximizes enjoyment, thereby diverting one from being drawn into a life that faces vulnerability and is rewarded by seriousness and meaning.

Someone seeking serious commitments and the lasting meaning they promise could enter the virtual world, but such a seeker would have to resist what is most seductive about the virtual world, viz., the promise of freedom from finitude. One would have freely to give up one's unrestrained freedom and make oneself vulnerable. Only then could one experience the excitement of bold transformation, or the grief and bliss of unconditional commitment. But then there would be no reason to spend a minute of one's life in an artificial world whose special attraction was its risk-free enjoyment.[20] There are plenty of opportunities for dedication with its concomitant dangers and rewards in the real world.

ARE ANY SOURCES OF MEANING NECESSARILY ABSENT FROM *SECOND LIFE*?

A serious philosophical question remains. Are there any rewarding ways of life not just discouraged but impossible in virtual worlds? That is, does an at least memorably meaningful life involve any crucial elements that may well be unprogrammable? As philosophers we will not be asking mereley about the limitations of current technology where a meanginful life is concerned, nor what people so far have used *Second Life* for or may use it for in the future. We are mainly interested not in actualities but possibilities; in this case (1) the necessary limitations of a certain model of human interaction dictated by a computer interface akin to the one in *Second Life*, and (2) the limitations, if any, on *all* human interactions in a virtual world. In keeping with the overall argument of this book, we might expect that, if there are such limitations, they will have to do with the importance of our real-world embodiment.

To answer these questions, where the meaning of life is concerned, we have to begin by noting that the most meaningful and rewarding kind of life we have discussed so far is an openness to a calling that, if answered, results in a life of enduring commitment. But Nietzsche first, and many postmodern thinkers since, have claimed that such an unconditionally committed life is rigid and restrictive and therefore less and less appealing,[21] while a life open to experiment and change has come to be seen as more and more attractive. The success of *Second Life* confirms this observation. But, as Kierkegaard points out, an experimental life lacks seriousness and focus. So the question arises whether our culture, or any culture, has practices that support a rewarding way of life that avoids both the narrow focus and immutability of traditional unconditional commitment as well as the hyper-flexibility and dispersion characteristic of life in our postmodern world.

In answer, Martin Heidegger has pointed to a familiar but now endangered species of practice that is more flexible than unconditional commitment but which, nonetheless, can provide focus, enchantment, and a memorable sort of meaning. Such practices can bring us in touch with a power that we cannot control and that calls forth and rewards our efforts – a power that we, therefore, recognize as sacred.

Heidegger has in mind practices that encourage local gatherings around things or events that set up local worlds. According to Heidegger, such local worlds bring out at their best those involved. Heidegger gives as an example drinking the local wine with friends, where a celebratory occasion, friendship, and a sense of being blessed can come together radiantly and forcefully. Albert Borgmann has usefully called the practices that support such local gatherings, *focal practices*.[22] The family meal when it acts as a focal practice requires the culinary and social skills of family members and draws fathers, mothers, husbands, wives, and children to come forth at their best. Such practices make family gatherings matter.

For people who experience such focal practices, many elements of the practice such as how and when to share a meal together can vary, but the basic focal practice itself is felt as an imperative, not a matter of choice. One does not simply choose the roles of family members. Nor does one simply choose the conventions of sharing a meal. These are the background on which all manifest options appear. Indeed, to do their work such practices *must* remain in the background. One reason we cannot program them is that we are so immersed in them that we cannot stand back from them and make them totally explicit.

For an example of a background practice that is taken for granted and can't be made explicit and programmed take distance standing. We are not aware that, when interacting with friends, colleagues, loved ones, and so forth we stand at what we feel to be the appropriate distance from them. If we thought about what distance to stand at, we wouldn't know how to do it. The sense of appropriate distance was passed on to us by our parents and peers who didn't know that they had the practice. They just felt uneasy and backed away when we stood too close and moved closer when they felt we were too far away, and now we do the same. Like many social skills, we mastered distance standing by our body conforming to other people's bodies.[23]

Anthropologists try to measure and codify the distance-standing practices in various cultures. There is even a field called Proximics dedicated to doing just this. But our distance-standing skill, like any skill, is endlessly flexible. We feel comfortable standing further away if the person we are interacting with has a cold, closer if there is a lot of noise in the background. In a library reading room or a church we speak more softly and stand closer, all these subtle discriminations and responses are further inflected by our relationship with the person involved.

So just how could such practices be introduced into the virtual world? The answer is surprising and important. The bodies of the users controlling the avatars bring them in. Experiments have shown that, without thinking about it, users tend to position their avatars in relation to each other at what would count as the appropriate distance in the real world.[24] We shall need to come back to this phenomenon in a moment.

In addition to basic background practices like standing appropriate distances from others, focal occasions require a shared mood and the sense that all who are present are sharing that mood. This sense of sharing creates a self-contained world. The best way to see this is to consider some famous representations of focal occasions from film and literature. Consider the dinner in the film *Babette's Feast*. At the beginning of the dinner, bickering among the guests over issues brought in from the past almost spoils the occasion by preventing it from becoming self-contained. But then, with the wine and good food, a mood of openness and care for others specific to the occasion descends, and when everyone senses that this mood is shared, the feast works as a self-contained world. Likewise, in Virginia Woolf's novel, *To the*

Lighthouse, Mrs Ramsey's dinner cannot come off as a successful occasion as long as a mood of political argument brought in from outside by the men persists. Only when a shared appropriate mood – in this case a mood of warmth and generosity – arises, and the guests sense that they are all sharing that mood, the event becomes a self-enclosed focal occasion.

A similar phenomenon occurs when there is a brilliant play at a baseball game and many in the crowd rise as one. What is so moving is not just that they are swept up in the same excitement; what is especially moving is that *each one* senses that they are *all* swept away by it. Indeed, the sense that the shared mood is shared is constitutive of the excitement. Again, it is what binds the participants together in a focal event and makes the occasion into a self-contained world.

When a focal event is working to the point where it has its particular integrity, one feels extraordinarily in tune with all that is happening, a special graceful ease takes over, and events seem to unfold on their own. This makes the moment an all-the-more enchanting and unforgettable gift. One feels grateful for receiving all that is brought out by this particular occasion, thus a reverential sentiment can arise. Such sentiments are frequently manifested in practices such as toasting or in wishing others could be joining in. An ancient practice for expressing such a sentiment was pouring a libation to the gods.

We have little current vocabulary for talking about our moods coming together to make an event come alive, but we know it is not in our power to make it happen.[25] How the power of moods is understood depends on the culture, but the understanding of moods as gifts from powers outside of our control is found in every culture, with the possible exception of ours.

A sense that we did not and could not make the occasion a centre of focal meaning by our own efforts, but rather that we were granted the special attunement required for such an occasion is what Heidegger wants to capture in his claim that for a focal event to work the divinities must be present. Describing a similar phenomenon – a baseball game where people are attuned to each other and sense that they are so attuned – Borgmann says:

> Given such attunement, banter and laughter flow naturally across strangers and unite them into a community. When reality and community conspire this way, divinity descends on the game.[26]

Much that gives life meaning is organized around such focal occasions. There are not only dinners and sporting events, but also celebrations such as weddings, graduations, and reunions, solemn commemorations such as memorials and funerals, as well as religious rituals such as Seder or the Eucharist. All these focal events depend for their success on the gift of a shared mood and the appreciation that it is shared. To determine whether this practice that helps make life worth living in the real world is reproducible in virtual worlds we must begin by considering to what extent moods can be experienced, communicated, and shared in *Second Life*.

In so far as philosophers have thought about moods at all, the usual approach until recently has been to think of them as *inner* mental states. On this Cartesian view, people are not really *in* a mood but moods are *in* people. A person's private moods are expressed (made outer) by his or her bodily movements, which can then be observed, interpreted, and responded to by another person's movements.

Given the mediation of the computer, the communication

of moods in *Second Life* is currently implemented the way Cartesians envisage the transmission of moods in the real world. If a resident in *Second Life* sitting at her computer experiencing a mood wants to communicate it to another resident, she must command her avatar to signal this private mood publicly by means of a preprogrammed gesture. The viewer then must interpret the gesture. If, thanks to his inner mental process, he succeeds in figuring out the mood of the sender from the gesture of the sender's avatar, he can then command his avatar to respond with an appropriate gesture. This way of understanding the communication of moods in *Second Life* makes manifest the clumsy character of the Cartesian account of our everyday communication of moods. But this Cartesian procedure does not at all capture the way moods are normally shared in the everyday world.

Stephenson, prophetic as usual, is onto this problem. He notes how important body language is in international negotiations and has Hiro observe:

> Businessmen . . . more or less ignore what is being said . . .
> They pay attention to the facial expressions and body
> language of the people they are talking to.[27]

But Stephenson doubts that programming body language would be sufficient to capture genuine emotional communication. He doesn't tell us the basis for his doubts; he simply has Hiro report that Juanita, the metaverse's master programmer who has done more than anyone else to program facial expressions and body language, does not believe her programs capture how people communicate their feelings. She thinks that there is something misguided in the whole programming approach. Hiro says:

> Juanita . . . has . . . decided that the whole thing is bogus. That
> no matter how good it is, the metaverse is distorting the way
> people talk to each other.[28]

It's hard to say what Juanita has in mind, but since Juanita is a master programmer and knows that programs can be improved without limit, whatever is lacking would have to be, not better programs, but something necessarily missing from how people currently communicate in virtual worlds — something that could not be fixed by programming more and more sophisticated gestures and facial expressions.

A comment from the official guide gives a hint of what is bogus about communication in the metaverse. Iris Ophelia, one of the residents of Second Life, while praising Second Life's attractions, admits:

> One of the biggest problems with the Internet since day one
> has been a lack of expression. Emoticons [smiley faces, etc.]
> help, but there's always an uncrossable line where
> expressions, tones, and body language lie . . .
>
> This whole world [of Second Life] has been created, with so
> much to see and do and experience, and yet there's so little
> genuine emotion. The crying gesture is used as a joke 90% of
> the time. If you were really crying, how could you convey it in
> Second Life? (207)

The question is, just what is missing? It seems that, given the Cartesian understanding of the communication of feelings, one would have to program a repertoire — a dictionary so to speak — of emotive gestures, and residents would have to choose which ones to use on each occasion. A certain conventional gesture of a person's avatar would be used to indicate being in a typical mood. The crying gesture is an extreme case. One might, to take a more everyday example, decide to use a gesture such as yawning to indicate one was bored.

But in the everyday world, moods are not normally experienced as essentially private and then communicated indirectly by using gestures. There are in fact two problems concerning communication of moods in *Second Life*. As already noted, to be programmed, the gestures used have to be generic while in the real world our communication is normally specific to each specific situation. Moreover, and more importantly, in our world the communication of our moods is *direct*, while in *Second Life* it is *indirect*. That is, in the real world our bodies *spontaneously* express our moods and others *directly* pick them up, while in *Second Life* one has *to select* an appropriate gesture and then *command* one's avatar to make that movement while the other person has to *figure out* what the gesture means. Thus the Cartesian model inserts an object/body – human or avatar – into the experience of everyday communication and thus distorts both the situation-specific moods we normally express and our spontaneous, direct, embodied, way of expressing them.

If stepping back and choosing a gesture were required to communicate our moods, communication would take us out of the flow of our immediate moods and transform them into self-conscious experiences, as if like an actor we needed to decide which bodily expressions to use. This is presumably why Juanita says that all emotional communication in the metaverse is bogus. Happily, in the real world people directly pick up and directly respond to each other's situation-specific moods. Indeed, genuine communication of moods seems to require the direct body-to-body interaction that in discussing the acquiring of distance-standing practices I called intercorporiality. As Merleau-Ponty puts the problem:

The sense of our gestures is not given, but grasped, that is, recaptured by an act on the spectator's part. The whole difficulty is to conceive this act clearly without confusing it with a cognitive operation. The communication or comprehension of gestures comes about through the reciprocity of my intentions and the gestures of others, of my gestures and intentions discernible in the conduct of other people. It is as if the other person's intention inhabited my body and mine his.[29]

Until recently, our direct communication of our feelings has been, indeed, mysterious, but recent work in neuroscience has cast a new light on the subject. Researchers have found brain cells, which they appropriately call mirror-neurons, that fire both when one makes a meaningful movement and when one sees another person make that movement.

As reported by Sandra Blakeslee in the *New York Times*:

The human brain has multiple mirror neuron systems that specialize in . . . understanding not just the actions of others but . . . the social meaning of their behavior and their emotions. [Giacomo] Rizzolatti says . . . "Mirror neurons allow us to grasp the minds of others not through conceptual reasoning but by feeling, not by thinking."[30]

Vittorio Gallese, the discoverer of mirror-neurons, provides more details:

When we observe actions performed by other individuals our motor system "resonates" along with that of the observed agent. Action observation both in humans and monkeys seems to imply a concurrent action simulation. This notion is corroborated by evidence coming from neurological patients.

Demented patients with "echopraxia" . . . show an impulsive tendency to imitate other people's movements. Imitation is performed immediately with the speed of a reflex action. Imitation concerns gestures that are commonly executed as well as those that are rare and even bizarre for the observing patient. It can be hypothesized that echopractic behavior represents a "release" of a covert action simulation present also in normal subjects, but normally inhibited in its expression. . . .[31]

Gallese notes that yawning is a normal case where the inhibition seems to be missing.

(Examples of) "contagious behavior" commonly experienced in our daily life, in which the observation of particular actions displayed by others leads to our repetition of them, [are] yawning and laughter.[32]

Moods are likewise contagious. No *interpretation* of someone's movements and no *selected* response movements are required. Of course, the direct communication caused by mirror-neurons only works if one is in the presence of a body enough like one's own. Cats' yawning doesn't make us yawn.

It's an empirical question whether an avatar's gestures can be made similar enough to ours to cause a direct response in the person controlling the avatar. But even if avatars could be programmed to make such realistic gestures that a person seeing the avatar on her computer would directly respond to it, she would still have to consciously command her avatar to make an appropriate canned response. So her response would still be doubly bogus, that is, not situation-specific and not direct. Indeed, indirectness is built into any model of communication that inserts two public object-bodies between two inner minds, whether the two interposed bodies are each person's own body as in the phenomenologically inadequate Cartesian model of everyday communication, or two avatar bodies as in the current Cartesian implementation of emotional communication in *Second Life*.

Instead of trying to explain how one's private inner states can be conveyed to others by means of one's public external body, Heidegger starts with the observation that moods are attunements and notes that attunements, unlike feelings and emotions, are normally public and directly shared. He describes the phenomenon:

> A human being who . . . is in good humor brings a lively atmosphere with them. . . . Or another . . . makes everything depressing and puts a damper on everything . . . What does this tell us? Attunements . . . in advance determine our being with one another. It seems as though an attunement is in each case already there, so to speak like an atmosphere in which we first immerse ourselves . . . and which then attunes us through and through.[33]

Heidegger implies that the traditional account of moods as private inner states misses the phenomenon of the contagion of moods. He asks:

> Do [moods] bring about an emotional experience which is then transmitted to others, in the manner of infectious germs? We do say that attunement or mood is infectious.[34]

And he further notes that most of the time and most basically people are directly attuned to each other by being always already attuned to a shared situation.[35] He writes:

> [Moods] are precisely a fundamental manner . . . of . . . being

> with one another . . . [a]nd precisely *those* attunements to
> which we pay no heed at all, the attunements we least
> observe, those attunements which attune us in such a way
> that we feel as though there is no attunement there at all, . . .
> – these attunements are the most powerful.[36]

Moods are powerful in that they are not under our control, and yet they determine what matters in our interactions with others and so govern our social behaviour.[37]

We therefore need to understand how people alone at their computers could be drawn into an already shared public mood in the virtual world.[38] It would seem that the current object-body-mediated model poses an insurmountable barrier to the genuine communication of moods in *Second Life*.[39] Phillip Rosedale, however, tells me that the programmers at Linden Lab are now working on just the sort of direct communication of one's feelings I would have thought impossible in *Second Life*. Linden Lab is developing software he says, that, if one has a webcam trained on one as one sits at one's computer, will enable the computer to pick up directly one's head and upper body movements and use them to control the movements of one's avatar. He says that "the technology exists today in every web camera that's out there to have it be the case that . . . if you're nodding or if you're making head movements, . . . – your avatar – [will make the same movements]".[40] So, your avatar could in principle directly manifest your feelings. This would be an important first step towards virtual intercorporiality!

There are problems, however. Although a camera can surely capture your posture, style, speed, and facial expressions, it is an open question how much of that information can be manifested by your avatar. The avatar's body, and especially

117 **Virtual embodiment**

its face, would have to be sufficiently human looking to reproduce the subtle movements that would be directly picked up by the camera. Whether the body language that the camera directly picked up could be reproduced in sufficient detail by one's avatar to communicate one's feelings directly to the viewer is an empirical question.[41]

If reproducing such subtle body movements were possible, people at their computers, already in a mood, might transfer their moods into their avatar's reactions without realizing they were doing so, just as they smuggle in background-standing practices. Capturing each person's movements and communicating them directly to his or her avatar might result in all the avatars getting in sync and so producing a contagious situational mood. Like an atmosphere, such a mood would be beyond the control of any one person and would draw in each new participant like a raindrop into a hurricane.[42] This is in principle possible but far beyond current technology.

Given the current Cartesian model, the best one can do is to direct one's avatar to go through the motions of being in a mood at a wedding, a funeral, a sporting event, or a family dinner but there would be no possibility of a contagious global atmosphere. Moods could only be experienced as private inner feelings communicated between isolated individuals by controlled body movements just as Cartesian philosophers have held. The spontaneity and specificity of shared attunements, and the sense that the shared attunements were shared, all of which go to make up a focal event, would necessarily be lacking. There could be no contagion, no excitement of being swept up into a shared atmosphere, no self-contained shared world, and no shared sense that something important and gratifying was happening. No divinity would descend and produce a memorable focal event.

To sum up: A focal event – perhaps the most meaningful experience available to us in our otherwise secular world – requires four capacities recognized by Heidegger and Merleau-Ponty that cannot be captured in the currently accepted Cartesian model:

1 *Intercorporiality*, i.e. the *direct* bodily expression and pick up of moods,
2 that the moods picked up be *shared*,
3 that those involved in a focal event sense that *the shared attunement is shared*, and
4 that those involved sense that they have *contributed* to their being *taken over by a power outside their control*.

As long as one is confined to the current Cartesian model of the communication of feelings, programming the contagion of moods is impossible, and so focal events are not possible in current virtual worlds.[43] However, we can begin to see that perhaps programmers at Linden Lab might generalize their webcam-using program and so discover how to smuggle people at their computers into the bodies of their avatars. After all, if programmers managed to program avatar bodies to make expressive movements sufficiently similar to ours, and if they could couple control of one's avatar directly to one's brain or body, they could perhaps draw on the mirror-neurons of the users to capture intercorporiality. Residents of *Second Life* could then be drawn into a shared mood and come to share that that shared mood was shared, and so bring focal practices into *Second Life*. Whether in fact focal events can be programmed, and if so how and when, are empirical questions.

CONCLUSION

We have seen that *Second Life* as currently conceived is subject to four philosophical objections. Existentialists would claim that indulging in a virtual life is the ultimate form of diversion to avoid facing the vulnerability of a real-world life. It would thus blind users to the anguish and joy of responding to a calling to face up to their finitude. Nietzscheans would see *Second Life* as a masquerade that offers cautious experimentation but misses the rewards of the sort of bold experimentation only possible in the risky real world. Kierkegaardians would say that the attraction of the safety of *Second Life* makes unconditional commitment unlikely. And finally, Heideggerians would point out that for a meaningful life one must be able to engage in focal events, and that that requires a sensitivity to the power of the shared moods that give mattering to our world, make possible focal events, and thus give meaning to our lives. But such sensitivity is impossible given the current Cartesian model of a concealed computer user deliberately controlling his public avatar.

Thus, as long as one works within the Cartesian framework of inner minds and object-bodies, a fundamental crosscultural ancient and modern way of making life worth living would inevitably be absent from virtual worlds such as *Second Life*. The idea that one could lead a memorably meaningful life in the kind of metaverse we currently can envisage would be a myth. For the time being, if we want to live life at its best, we will have to embrace our embodied involvement in the risky, moody, real world.

We have now seen that our body, including our emotions and moods play a crucial role in our being able to make sense of things so as to see what is relevant, our ability to let things matter to us and so to acquire skills, our sense of the reality of things, our trust in other people, and, our capacity for making the unconditional commitments that give a fixed meaning to our lives, and finally the capacity to cultivate the intercorporiality that makes possible meaningful focal events. It would be a serious mistake to think we could do without these embodied capacities – to rejoice that the World Wide Web offers us the chance to become more and more disembodied, detached, ubiquitous minds leaving our situated, vulnerable bodies behind. The increased disembodiment of information leads to difficult trade-offs.

In Chapter 1 we saw that up to 1999, as the Web grew alarmingly, people were faced with a painful trade-off between high speed statistical syntactic search of meaningless hyperlinks, and slow old-fashioned human judgments of meaningful connections among pieces of information. This led to desperate attempts, in the face of repeated failures, to formalize intelligence and natural language. But now in the new millennium, thanks to Google and Wikipedia, we can stop wasting time and money on AI and natural language processing and enjoy the best of both worlds – high speed syntactic search of billions of hyper-linked Webpages for what is important to users, and human judgment as to how best to organize vast amounts of information about the world so as to preserve and bring out its meaningful connections.

In the other four chapters, however, the trade-offs are more complicated. The two options are not equal; one side of the trade-off is superior to the other. One might call these asymmetrical trade-offs.

In Chapter 2 we saw that, as far as education is concerned, the Net can be useful in supplying the facts and rules as well as the drill and practice required by a beginner. It seems, however, that the involvement and risk that come from making interpretations that can be mistaken and learning from one's mistakes are necessary if one is to acquire expertise. Such involvement is absent if one is just sitting alone in front of one's computer screen looking at a lecture downloaded from the Web. There is more involvement in an on-line interactive lecture on the Web, but the sense of taking a risk and accepting approval or criticism in front of others is much reduced, and, therefore, so is the involvement. Such lectures are, therefore, not likely to produce more than competence. Only in a classroom where the teacher and learner sense that they are taking risks in each other's presence, and each can count on criticism from the other, are the conditions present that promote acquiring proficiency, and only by acting in the real world can one acquire expertise. As for the apprenticeship necessary to becoming a master, it is only possible where the learner sees the day-to-day responses of a master and learns to imitate her style.

Thus, we saw in Chapter 2 that, in considering distance education, one has to choose between economy and efficacy, and that, while administrators and legislators tend to prefer the "maximum throughput" even if it can only produce competence, most teachers, parents, and students, if they can afford it, would prefer the shared involvement that produces proficiency, and the real-world experience and mentoring that makes possible the acquisition of expertise and mastery.

Where our sense of reality is concerned, the trade-off is differently asymmetrical. The relation of presence to telepresence is not a question of the advantages and disadvantages of each, and so of choosing one over the other. Rather, telepresence presupposes presence. Here, the asymmetry is one of dependence. Thus, I argued that telepresence, both of objects and people, is parasitical on a robust sense of the presence of the real correlative with the body's set to cope with things and people.

Where meaning is concerned, again the trade-off is asymmetrical. This time, one side is positive and the other negative. If we remain the kind of beings that Kierkegaard understood us to be, we will despair if all meaningful distinctions are levelled, and since Judeo-Christian meaningful distinctions require commitment and vulnerability, which require our embodied finitude, we should have no trouble in choosing between disembodied nihilism and embodied meaningful differences.

We may lament the risks endemic to an embodied world where we are embedded with objects and others in local situations, but the idea of living in boundless virtual worlds, where everyone is telepresent to everyone and everything, levels all significant differences and offers no support for being drawn into local meaningful events.

But isn't all this just to say that we can see what the Web can't do for us, but there may be great things it can do that we can't yet even imagine. After all, in the "Phaedo" Plato famously objected to the introduction of writing as opposed to speech, because, as he pointed out, writing reduces the richness of communication since it makes it impossible to read the speaker's tone and bodily posture. Furthermore, he saw that, if agreements could be made at a distance, they would not be as binding as agreements sealed by the spoken word. He also thought that people would lose their ability to remember important events.

Of course, all of that was true, but Plato couldn't foresee that, thanks to writing, we would gain a wider range of communication, new ways of making contracts at a distance, and a whole new cultural memory. If he could have foreseen all this, he might well have had a more positive view of the trade-offs involved.

No doubt the Internet, like the car, will have huge consequences both for good and ill that we cannot foresee. Nonetheless, there are two important differences between my argument and Plato's. I don't know what the claims for the value of writing among Plato's contemporaries were, but I have been arguing that the positive claims for the value of the Internet offered by our contemporaries are mostly hype. Whatever the long-range value of the Net turns out to be, it won't be the quality of information it offers, the democratic distance learning it makes possible, the presence of the Net user to all of reality, and the possibility of an experimental life full of meaning yet safe from world collapse.

More importantly, if my arguments are right, the Net differs dramatically from writing as to what an uncritical use of it could lead us to lose. It's unlikely that any of Plato's contemporaries were proposing that everyone would be better off the more they gave up talking and lived their lives through writing, whereas we are being told by groups like the

Extropians that, the more we can give up our bodies and live in cyberspace, the better off we will be. My answer is that, if we managed to live our lives in cyberbia, we would lose a lot more than the face-to-face conversations, verbal promises, and memory power Plato saw were endangered by writing. We would lose our only reliable way of finding relevant information, the capacity for skill acquisition, a sense of reality, and the possibility of leading meaningful lives – the last three of which are constitutive of us as human beings. Indeed, they are so definitive of who we are that nothing new and unexpected could possibly make up for our losing them.

But we would, of course, still like to know what the Web is good for and what it is not, so we can use it for what it does well. How then can we profit from the Web in each of the above areas? Obviously, we need to foster a symbiosis in which we use our embodied positive powers, to find what is relevant, learn skills through involvement, get a grip on reality, make the risky commitments, respond to the shared moods that give life meaning, and foster as much direct access in cyberspace to our avatar bodies as possible, while letting the Web contribute its amazing capacity to store and access astronomical amounts of information, to connect us to others, to enable us to be observers of far-away places, and to experiment without risk with other worlds and selves. In place of a summary of what has already been said, then, I'd like to offer a few examples of how this symbiosis might work.

1 SEARCH, RELEVANCE AND RETRIEVAL

In 1999 responsible observers claimed that the Web was growing so fast that syntactic search must fail and that people must fall back on human judgment. Here is a typical summary of the situation at the turn of the millennium:

> When search engines first appeared, they were hailed for accomplishing two things that could not be done by people on any large scale: Search engines used software agents to find and index sites almost as soon as they appeared. And they could almost instantaneously match a far-flung Web page with a single keyword typed into a beckoning search box.
>
> But the promise of automation has been tempered by the Web's success. There are now more than one billion Web pages, and according to some experts' calculations, the number has been doubling once every eight months. . . . To cope, many search engineers have concluded that simply indexing more pages is not the answer. Instead, they have decided to rely on the one resource that was once considered a cop-out: human judgment.[1]

But it now looks like Google has shown that there is no need for old-fashioned human judgment and the libraries and encyclopedias it enabled people to organize and search.

But does the success of Google show that the pessimism of the late 1990s was simply wrong? Was Don Swanson mistaken when he said, "Machines cannot recognize meaning and so in principle cannot duplicate what human judgment can bring to the process of indexing and classifying documents"? Have computers been programmed to show human-like judgment? No, Google has demonstrated that there is a syntactic way to use human judgment to compute importance and even a syntactic way to compute relevance. Thus, using the Google approach search gets better the more Websites there are to be searched.[2]

But surprisingly, it has also turned out that the *New York Times* reporter cited above was right, but for the wrong reasons. The growing size of the WWW was not the problem.

Google has solved the daunting problem of searching billions of Websites for information organized by hyper-links. Nonetheless, there has been a return to human beings organizing a vast body of information to fit the interests of other human beings. Alongside Google and its mining of horizontal hyperlinks, human volunteers have appeared who are experts in some specific domain and who have common sense like the encyclopedists of old. These experts use their experience and common sense to organize material according to its meaning. The result is Wikipedia, a human-edited and maintained on-line encyclopedia organized in the old vertical way.

Moreover, Gordon Rios notes that, "The biggest story is the increase in usage of Wikipedia. It has, since its inception in 2001, been closing the gap with Google as to their percent of daily page views. Often Google's best results come by just pointing you to the right page on Wikipedia."

Both methods of search are valuable. Sometimes syntactic

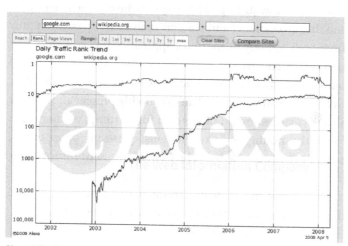

Chart 1 Traffic history graph for Google and Wikipedia

search works best and sometimes old-fashioned semantic search does a better job.[3]

This is a beautiful example of human beings doing what computers can't do – namely exercising judgment in organizing information so as to bring out how the bits of information are *relevant* to each other, while at the same time programming computers to do what human beings can't do – namely, search huge amounts of hyper-linked data, without understanding it, for what human users find important.

Pessimism is no longer the order of the day. The future of search on the web is bright both for computer users using Google's capacity to mine meaning out of intrinsically meaningless hyper-links, and also for the judgment calls of human encylopedists and librarians organizing information in a vertical way that makes sense to human beings on the background of their shared embodied human form of life.

2 DISTANCE LEARNING AND THE PODCAST WORLD

Granted that acquiring skills requires involvement and risk, and that professional and cultural skills can be passed on only from body to body by means of apprenticeship, still, in education, there are many ways of combining the advantages of old-fashioned lecture/discussions with the power of the Web. Given a class in which students are bodily present and there is already a shared mood of concern for learning, teachers have found that putting their assignments, questions, paper topics, etc., on a course Website helps students stay informed as to what is going on in the course. Teachers can also pose questions that the students can discuss in a newsgroup, and they can intervene in the discussion when necessary to clarify issues raised by the students.

In addition, I've found that it's useful to put my actual

lectures in MP3 format on my course Website, so that students who have to miss class can listen later from their dorms, and students writing papers can review lectures that went by too fast for them to follow. Of course, if one discusses films in one's courses, as I do, one can also include on the course Website film clips of the scene being discussed cued to the audio.

I've also gone a step further and arranged to Webcast, i.e. post in video, one of my courses so that students can watch the course from their dorms rather than sit on the floor in a crowded classroom. One might wonder why, in such a case, students would bother to come to class at all, but most students must be getting something special out of being bodily present at the lectures – sharing the mood in the room and making risky suggestions in class discussion – since, although they can now watch the lectures on their computers any time that is convenient, class attendance has barely been affected, except on rainy days, when attendance drops by about 30 per cent. This suggests that presence in class is felt to be such a positive experience that most students will slog through bad weather to attend, but that the Webcast offers enough so that those who cherish their comfort can make do with distance learning – it is certainly better than nothing.

Students who watched my lectures from their dorms said that they found the archived Webcasts helpful *once they had been to the lecture*. Thereafter, they could replay the Webcast, stopping the video to go over difficult points. But they felt that there was something about being present in the room with the lecturer and the other students that give them a sense of "interconnectedness" that they would not want to do without. They also felt that the presence of the lecturer focused them on what was important in the material being presented. Yet they also said, to my surprise, that they preferred the *audio* version of my lectures. They said they found the moving image merely distracting.

But, as of this second edition, things have taken a surprising turn. Over the past five years most elite colleges and universities have abandoned their distance learning projects. Instead of trying to replace the classroom and democratize high education these universities have remained as elite as ever but they have reached out to all lovers of knowledge by making the disembodied aspect of their courses available to everyone all over the world. An Associated Press article that has received wide coverage tells the story:

> When the Internet emerged, experts predicted it would revolutionize higher education, cutting its tether to a college campus. Technology could help solve one of the fundamental challenges of the 21st century: providing a mass population with higher education at a time when a college degree was increasingly essential for economic success.
>
> Today, the Internet has indeed transformed higher education. A multibillion-dollar industry, both for-profit and nonprofit, has sprung up offering online training and degrees. Figures from the Sloan Consortium, an online learning group, report about 3.5 million students are signed up for at least one online course – or about 20 percent of all students at degree-granting institutions. But it hasn't been as clear what role – if any – elite universities would play in what experts call the "massification" of higher education. Their finances are based on prestige, which means turning students away, not enrolling more. How could they teach the masses without diminishing the value of their degree?[4]

MIT pioneered an answer:

An MIT initiative called "OpenCourseWare" makes virtually all the school's courses available online for free – lecture notes, readings, tests and often video lectures. MIT's 2001 debut of OpenCourseWare epitomized a key insight: Elite universities can separate their credential from their teaching – and give at least part of their teaching away as a public service. They aren't diminishing their reputations at all. In fact, they are expanding their reach and reputation.

MIT's initiative is the largest, but the trend is spreading. More than 100 universities worldwide, including Johns Hopkins, Tufts and Notre Dame, have joined MIT in a consortium of schools promoting their own open courseware. . . . This month, Yale announced it would make material from seven popular courses available online, with 30 more to follow.

As with many technology trends, new services and platforms are driving change.

Some universities started putting lectures on the iTunes store in the form of podcasts, which are free video or audio recordings that anyone can download to their computer or iPod. The downloads have surged since May, when Apple began featuring lessons on the iTunes home page under the heading iTunesU. For example, the 86 courses UC Berkeley offers are now being downloaded 50,000 times a week, up from 15,000 before Apple's promotion.[5]

This approach seems to me the right way to go. It uses the Web to make the disembodied part of elite education available to everyone, while not claiming, in fact obviously denying, that this passive, disembodied form of learning can replace learning in the risky presence of a professor and fellow students. So I've made all my current courses available on iTunesU. A recent *LA Times* front-page story,[6] picked up and expanded by ABC News,[7] reports on my involvement:

Baxter Wood is one of Hubert Dreyfus' most devoted students. During lectures on existentialism, Wood hangs on every word, savoring the moments when the 78-year-old philosophy professor pauses to consider a student's comment. . . . But Wood is not sitting in a lecture hall on the UC Berkeley campus, nor has he met Dreyfus. He is in the cab of his 18-wheel big rig, hauling dog food from Ohio to the West Coast or flat-screen TVs from Los Angeles to points east. The 61-year-old trucker from El Paso eavesdrops on the lectures by downloading them for free from Apple Inc.'s iTunes store . . . then piping them through his cabin's speakers. He hits pause as he approaches cities so he can focus more on traffic than on what Nietzsche meant when he said God was dead, then shifts his attention back to the classroom. "I'm really in two places at once," he said. "The sound of chalk on the chalkboard makes it so real."

By making hundreds of lectures from elite academic institutions available online for free, Apple is reinvigorating the minds of people who have been estranged from the world of ideas. The universities want to promote themselves to parents and prospective students, as well as strengthen ties with alumni. Some also see their mission as sharing the ivory tower's intellectual riches with the rest of the world. . . . These unofficial students, invisible to their instructors, won't earn degrees for listening. Some professors won't even respond to their correspondence. But they relish the explosion of free lectures. Retirees in Long Beach and Weaverville, Calif., halibut fishermen in Alaska, data entry clerks in London, casting agents in New York – all separated from the classroom by age, distance or circumstance – are learning from some of the world's top scholars. . . .[8]

This is the Internet at its best, providing a disembodied but nonetheless much appreciated form of education that would have been impossible without it.

3 TELEPRESENCE

I've argued that telepresence can never give us a sense of the risky reality of far-away things even when I can act at a distance and can see the results of my actions as in controlling the robot in the telegarden, nor can it convey a sense of trust of distant human beings. It therefore seems a waste of effort to try and make telepresence do the job of bodily presence by adding feeling, smells, etc. Still, as we have seen, there is a place for teleconferencing when people already know and trust each other. And, of course, telepresence is still indispensable in those areas for which it was developed, such as dealing with things where bodily presence is too big, too small, too risky, etc., as in repairing nuclear reactors and exploring unliveable planets. These possibilities pre-date the World Wide Web, but the Web can expand our perceptions and active intervention to the far corners of the universe. It is estimated that there are now over 15,000 Webcams in operation, and, through them, one can see the traffic or the weather at any time almost anywhere in the world. As long as we continue to appreciate our bodies and don't lose our engineering expertise by substituting distance learning for lectures and apprenticeship, our minds can, indeed, expand to more and more of the universe. We can look forward to improved versions of vehicles like the Mars rover that will explore distant planets with millions of earthbound televiewers on board.

Perhaps, the televiewers will even be able to guide these remote explorations. The use of robots along with Internet-mediated telepresence offers the attractive possibility of each of us being able to control distant representatives of ourselves that are extensions of our eyes and ears. We could then take part in situations too dangerous for us to explore in person, by, for example, walking into a nuclear reactor, or we could simply be present in situations that we were unable to attend, such as taking part in the Oscar awards ceremony while shooting on a set in France.

After forty years of being told that a household robot is just around the corner, one would think that such a robot slave who could represent us in dangerous and far-away places would be easy to build, and now, thanks to the possibility of telepresence, easy to control. After all, thanks to Ken Goldberg's Telegarden, one already can control a robot arm so as to plant and water a seed in Linz, Austria. Such uses of telepresence mediated by robots are sure to grow.

Sadly, however, reality lags far behind predictions. At an international meeting of robot makers at MIT, all but a few fanatics agreed that humanoid robots would, for a long time, remain science fiction. A *New York Times* reporter filed the following report:

> Last month's organizers of the Humanoids 2000 conference surveyed some of the participants about possible social implications of their work. On a scale of 0, for highly unlikely, to 5, for highly likely, the robotics researchers rated the possibility that robots "will be the next step in evolution and will eventually displace human beings" a zero. "They are much less euphoric than other people, say, movie producers", said Dr. Alois Knoll . . . one of the organizers of the

On the Internet

conference. . . . Dr. Knoll listed the limitations of present-day robots: "We don't have the mechanical dexterity. We don't have the power supply. We don't have the brains. We don't have the emotions. We don't have the autonomy in general . . . to even come close to humans."[9]

But not to worry, Ken Goldberg and his co-workers have suggested a solution that is now being explored by the MIT Media Lab and discussed seriously in Silicon Valley. Those involved realize that robots will, for a long time, be too clumsy to be our representatives, so they propose that we recruit actors to do the job. They are therefore working on how a person or a group of people could teleguide a Tele-Actor wearing the Webcams and microphones that would enable the controller to be telepresent at far-away events. The Tele-Actor, impersonating a robot, would wear goggles with lights around the edges that would signal to him or her which way to turn and how fast to move, etc. as the controller teleguided "it" to take part, for example, in a far-away award ceremony.

Fortune Magazine published the following report on the Media Lab project under the title "Being-There":

Send a Tele-Actor out to a location, and you see what it sees and hear what it hears. Multiple participants can log on, all sharing the same viewpoint, all helping to direct the action. "It lets anyone tap into a remote experience – a sports event, a conference, maybe even a place too dangerous for most people, like a war zone", says Ken Goldberg. . . . Goldberg created the idea with a team of colleagues as part of their experiments in "telepresence", which uses technology to break down distance. As bandwidth improves and camera tech gets cheaper, they see Tele-Actors becoming common.[10]

By proposing an ingenious end-run around the failures of AI and the setbacks of humanoid robot research, the Media Lab has succeeded in once again illustrating a disturbing tendency of computer enthusiasts. Computers exhibit the possibility of augmenting human capacities such as memory and calculating ability, but it turns out they lack other abilities such as intelligence and the ability to move their body in adaptive and coordinated ways. So, to take advantage of the possibility of telepresence provided by the Internet, since robots can't be programmed to behave like people, people will have to learn to behave like robots.

When you are guiding a Tele-Actor will you feel that you are bodily present at the scene relayed by the robot? Probably not. There will be no risk, no proprioception, no sense of directly causing the movement of the Tele-Actor (see Chapter 5). There will be, however, a sense of your controlling what you see and hear and that may be enough telepresence to make the proposal interesting.

4 REFLECTION VS COMMITMENT

Some of the most vexing questions arise over whether the World Wide Web is improving or diminishing the quality of our lives. We've seen that two surveys suggest that living through the Net leads to isolation, and one of these surveys finds, in addition, that use of the Net leads to loneliness and depression.

Yet a recent National Public Radio survey showed that people felt just the opposite of the ill-effects found in the Carnegie-Mellon and Stanford studies. I quote from the NPR Website:

A new poll by National Public Radio, the Kaiser Family

Foundation, and Harvard's Kennedy School of Government shows that people overwhelmingly think that computers and the Internet have made Americans' lives better. Nearly 9 in 10 say computers have made life better for Americans, and more than 7 in 10 say the Internet has made life better.[11]

Yet the poll also showed that "more than half [of those polled] say computers have led people to spend less time with their families and friends". This shows, I think, that not only are we transformed by the way we use our tools; we are not aware of how we are being transformed, so we need all the more to try to make explicit what the Net is doing for us and what it is doing to us in the process.

I've suggested that, where meaning is concerned, what the Net is doing to us is, in fact, making our lives worse rather than better. Living one's life on the Web is attractive because it eliminates vulnerability and commitment but, if Kierkegaard is right, this lack of passion necessarily eliminates meaning as well.

It should thus be clear that tools are not neutral, and that using the Net diminishes one's involvement in the physical and social world. This, in turn, diminishes one's sense of reality and of the meaning in one's life. Indeed, it seems that, the more we use the Net, the more it will tend to draw us into the unreal, virtual worlds populated by those who want to flee all the ills that flesh is heir to.

If, however, one is already committed to a real-world cause, the World Wide Web can increase one's power to act, both by providing relevant information, and by putting committed people in touch with other people who share their cause and who are ready to risk their time and money, and perhaps even their lives, in pursuing their shared end. The landmine treaty, for example, was hammered out and promoted largely thanks to the fact that the Web is international and has no gatekeepers.

But, the risk posed by the ambiguous similarity of social cyberspaces to communities in the embodied social world comes out clearly in the second edition of Howard Rheingold's influential book, *The Virtual Community*.[12] In his new chapter, "Rethinking Communities", Rheingold responsibly discusses a tangle of issues surrounding the advantages and disadvantages of many–one interactions in cyberspace. Unfortunately, his analysis is marred by his failure to distinguish the various forms such Internet communities can take.

To begin with, Rheingold defends his conviction that cybercommunities could improve democracy. "The most serious critique of this book", he says, "is the challenge to my claim that many–one-discussions could contribute to the health of democracy by making possible better communications among citizens."[13] He then goes on to develop the claim made in the first edition that the Net "might help revitalize the public sphere", indeed, that "the vision of a citizen-designed, citizen-controlled worldwide communications Network is a version of technological utopianism that could be called the vision of 'the electronic agora' ". "In the original democracy, Athens", he explains, "the agora was the marketplace, and more – it was where citizens met to talk, gossip, argue, size each other up, find the weak spots in political ideas by debating about them."[14]

But the vision of a *worldwide* electronic agora precisely misses the Kierkegaardian point that the people talking to each other in the Athenian agora were members of a direct democracy who were directly affected by the issues they were discussing, and, most importantly, the point of the discussion was for them to *take the responsibility and risk of voting publicly* on the questions they were debating. For Kierkegaard, a worldwide electronic agora is an oxymoron. The Athenian agora is precisely the opposite of the public sphere, where anonymous electronic kibitzers from all over the world, who risk nothing, come together to announce and defend their opinions. As an extension to the deracinated public sphere, the electronic agora would be a grave danger to real political community. Kierkegaard enables us to see that the problem is not that Rheingold's "electronic agora" is too utopian; it is not an agora at all, but a nowhere place for anonymous nowhere people. As such, it is dangerously distopian.

The discussion is blurred by the fact that Rheingold does not distinguish the negative influence of the contribution of the Net to the public sphere from two positive ways in which the symbiosis of embodied individuals and the disembodied Internet allows people to leap out of the prison of endless reflection: the *aesthetic possibilities* of virtual commitments, on the one hand, and the *ethical actuality* of committed action, on the other.

Virtual communities constitute an interesting leap into the aesthetic sphere of existence. Such communities are in a certain way the antithesis of the public sphere since passionate commitments are encouraged, not frowned upon, and the issues debated are of crucial concern to the virtual community. Kierkegaard agrees that people in the aesthetic sphere of existence are involved in each other's emotional lives. But what is essential to him is that, although the aesthetic person lives in a world of intense feeling and lively communication, all the drama is like a game in that it has no real-world consequences and there is no real-world risk. Individuals can enter or leave a virtual community much more easily than they can move out of a town they dislike. As we saw, Kierkegaard says that the aesthetic sphere turns existence into a play.

In his revised edition, Rheingold frankly faces the danger "that virtual communities might be bogus substitutes for true civic engagement".[15] And he acknowledges that:

> most of what needs to be done has to be done face to face, person to person – civic engagement means dealing with your neighbors in the world where your body lives. . . . Discourse among informed citizens can be improved, revived, restored to some degree of influence – but only if a sufficient number of people learn how to use communication tools properly, and apply them to real-world political problem-solving.[16]

One could conclude, and Rheingold might well agree, that, as a game, involvement in virtual communities is not a threat to political engagement in one's actual community. But it becomes harmful if, as in the case of *Second Life*, its risk-free nature makes it more attractive than the dangerous real world, and so drains off the time and energy that citizens could have given to actual community concerns.

So, in his new chapter, Rheingold's emphasis shifts to the role the Internet can play in bringing together people with concrete problems and enabling them to act more effectively. Thus, he proposes "experimenting with different tools for civic involvement".[17] But his defense of such Internet interest groups is presented as a defense of the public sphere, so that the important distinction between detached and anonymous talk and involved responsible action is lost. Rheingold's impressive list of Internet groups that foster concrete commitments – such as a group called Cap-Advantage that provides "Tools for Online Grassroots Advocacy and

Mobilization" – also includes free-floating public sphere groups like Freedom Forum, which he describes as "a non-partisan international foundation dedicated to free press, free speech and free spirit for all people".[18]

If in reading Rheingold's book one bears in mind Kierkegaard's threefold distinction between the public sphere with its reflective detachment from local issues, the aesthetic sphere with its risk-free simulation of the serious concerns of the real world, and the ethical sphere with its local political commitments, one can then be grateful to Rheingold for laying out the impressive spectrum of what the World Wide Web in symbiosis with local committed individuals can provide.

5 ACTING AND LEARNING THROUGH AVATARS

Besides podcasting, thanks to virtual worlds such as *Second Life* there are further new opportunities for improving distance learning using the technology of three-dimensional virtual worlds. After the *LA Times* article about podcasting, many podcast listeners e-mailed me their regret that they couldn't meet with other listeners each week to discuss the points made in my lecture. This led me to investigate organizing a discussion section in *Second Life*. I learned from the *Chronicle of Higher Education* that "more than 150 colleges in the United States and 13 other countries have a presence in *Second Life*". Furthermore, "although some faculty and staff members are skeptical of the digital world's value the number of virtual campuses keep growing. Indeed, Professors use *Second Life* to hold distance-education classes, saying that communication among students actually gets livelier when they assume digital personae."[19]

It seems to me unlikely, however, that the relatively dis-embodied students each alone in front of his or her computer and represented in the virtual classroom by an avatar could become as involved in the risky process of shared learning as can embodied students learning together in a classroom in the real world. If discussions are more lively when the students meet as digital personages, it seems to me that this must be because they were only half alive before. Perhaps in such a case, thanks to the novelty of the virtual world, they were at least somewhat more alive in *Second Life* than in their boring everyday real-world classroom.[20] Since the students were all on the same campus and so could be bodily present with a lively teacher, it's hard to see how being half-embodied in unexpressive avatars could be an improvement.

But, given that my podcast audience was spread all over the world and couldn't be in the presence of each other, the idea of a discussion section in *Second Life*, although only second best, nonetheless made sense. So, in response to requests from listeners worldwide who wanted to discuss with other listeners the lectures they were all hearing, as well as my own desire to check out how teaching would work in a disembodied virtual world, I decided to try teaching in *Second Life*.

Thanks to a computer-savvy podcast listener who volunteered to take care of the administrative and technological details, we set up a virtual discussion section in a beautiful virtual classroom overlooking the sea. There, my avatar, a young redheaded Farnsworth Roux, sat down with ten other avatars, all young and handsome like mine, and discussed Kierkegaard and Heidegger for over an hour.

Everyone who e-mailed me afterwards considered the discussion section an unqualified success. But I was frustrated by the fact that each of us had to be represented by an avatar. We couldn't really see each other. I couldn't see if the group around the table was gripped by the discussion or was bored and restless. I couldn't look the student who was speaking in the eyes to estimate his or her involvement. Moreover, there was no shared mood in the classroom that I could consciously and unconsciously shift and intensify. The whole situation struck me as Cartesian, in the sense that only the minds of those who where "present" were involved. I could understand the intellectual content of what each participant said but not how he or she was related to what they were saying.

So I asked the group, couldn't this meeting equally well have taken place by way of a conference call? One participant pointed out that the fact that they could raise the hands of their avatars when they wanted to speak, and that I could then call on them one by one, gave order to the conversation. Another pointed out that feeling that you are talking to some-one sitting opposite you or next to you or at the end of the table gave a sense of direction to the discussion.

I came away thinking that bodily presence offers such a rich educational environment that, if the students are on the same campus, it would be folly to give it up. But that, given the interest and enthusiasm of the podcast audience from all over the world, meeting as avatars to discuss the material is an educational possibility to be enthusiastically embraced not ignored or denigrated.

In sum, as long as we continue to affirm our bodies, the Net can be useful to us if we resist its tendency to offer the worst of a series of asymmetric trade-offs: economy over intensity in education, risk-free disembodied telepresence vs risky embodied interacting, the virtual over the real in our relation to things and people, detachment and anonymity over commitment in our on-line lives, and safe experimenta-tion offered by avatars over the bold experimentation offered by real bodies. In short, in using the Internet,we have to remember that our culture has already fallen twice, first for the Platonic and then for the Christian temptation to try to get rid of our vulnerable bodies – an attempt that has ended in nihilism. This time around, we must resist this temptation and affirm our bodies, not in spite of their finitude and vulnerability, but because, without our bodies, as Nietzsche saw, we would literally be nothing. As Nietzsche has Zarathustra say: "I want to speak to the despisers of the body. I would not have them learn and teach differently, but merely say farewell to their own bodies – and thus become silent."[21]

Notes

INTRODUCTION

1 The Extropians are very far out, but the same ideas show up in supposedly serious books such as Hans Moravec's *Mind Children*, Cambridge,
MA, Harvard University Press, 1988.

2 A tendency that Martin Heidegger claims is definitive of our modern
understanding of what it is to be anything at all. See Martin Heidegger,
"The Question Concerning Technology", in *The Question Concerning
Technology*, New York, Harper and Row, 1977.

3 Not that there haven't been real innovations. New ways of linking
information have transformed libraries; course Websites in colleges
and universities have made it possible for students to hear lectures and
engage in discussions without leaving their rooms; telerobotics has
made it possible to control a vehicle on Mars and one day millions
of spectators will no doubt be able to look out of such a vehicle as it
moves across the Mars surface; and e-mail has opened up surprising
new possibilities, from political dissidents working together for reform,
to proud grandparents sending their friends the latest digital photos of
their grandchildren. But all these surprising new developments are
minor compared to what has been predicted.

4 A. Harmon, "Researchers Find Sad, Lonely World in Cyberspace", *The
New York Times*, August 30, 1998. Harmon continues:

> Those participants who were lonelier and more depressed at the
> start of the two-year study, as determined by a standard question-
> naire administered to all subjects, were not more likely to use the
> Internet. Instead, Internet use itself appeared to cause a decline in
> psychological well being, the researchers said.

5 R. Kraut, M. Patterson, V. Lundmark, S. Kiesler, T. Mukophadhyay and W. Scherlis, "Internet Paradox: A Social Technology that Reduces Social Involvement and Psychological Well-being?" *American Psychologist*, 1998, vol. 53, no. 9, pp. 1017–31.

6 Ibid. It seems that lack of physical presence can lead to a kind of moral isolation too. When Larry Froistad confessed to his e-mail support group that he had murdered his daughter, the members of the group offered him sympathy; only one, Ms. De Carlo, felt they should turn him over to the police. See, "On-Line Thoughts on Off-Line Killing" by Amy Harmon, *The New York Times*, April 30, 1998. "It seemed to Ms. De Carlo that the nature of on-line communication – which creates a psychological as well as physical distance between participants – was causing her friends to forget their off-line responsibilities to bring a confessed murderer to justice."

7 J. P. Barlow, "A Declaration of the Independence of Cyberspace", Davos, Switzerland, February 8, 1996.

8 Moravec, *Mind Children*.

9 R. Kurzweil, *The Age of Spiritual Machines*, New York, Penguin, 2000.

10 Esther Dyson, George Gilder, George Keyworth, and Alvin Toffler, "Cyberspace and the American Dream: A Magna Carta for the Knowledge Age. Release 1.2" – August 1994, Washington, DC, The Progress and Freedom Foundation. http://www.pff.org/issues-pubs/futureinsights/fi1.2magnacarta.html.

11 Plato, "Gorgias", 492e7–493a5. Socrates says: "I once heard one of our wise men say that we are now dead, and that our body (*soma*) is a tomb (*sema*)."

12 Plato, "Phaedo", *The Last Days of Socrates*, Baltimore, MD, Penguin, 1954, p. 84.

13 F. Nietzsche, *Thus Spake Zarathustra*, trans. W. Kaufmann, New York, Viking Press, 1966, p. 35.

14 Ibid., p. 34.

ONE THE HYPE ABOUT HYPERLINKS

1 National Public Radio, "The Future of Computing", *Talk of the Nation, Science Friday*, July 7, 2000.

2 Fallows, D., «Search Engine Users», Pew Foundation, URL http://www.pewinternet.org/pdfs/PIP_Searchengine_users.pdf, 2005.12.

3 S. Lawrence and C. L. Giles, NEC Research Institute, "Searching the World Wide Web", *Science*, 280, April 3, 1998, p. 98. Moreover, the size isn't just the number of Websites or pages; the number of hyperlinks embedded in the Web pages is even larger.

4 There has been some interesting litigation of late trying to stop this "free-linking" of anything to anything, in which parties have sued others who made links to the plaintiff's Web page. Of course, this is a fraction of a fraction of a per cent, and is unlikely to have any significant effect on the way the Web is run which has been called a "loose ad-hocracy". It no doubt just reflects the dying gasp of the old guard who would like to place at least *some* limits on the eventual linking of everything to everything.

5 The Dewey decimal system was organized in this way. It did not even allow the same item to be filed under two different categories, but now librarians have more leeway and file the same information under several different headings. For example, Philosophy of Religion would presumably be filed under Philosophy and Religion. Still, however, there is an agreed-upon hierarchical taxonomy.

6 What people now refer to as the modern subject came into being in the early seventeenth century as – thanks to Luther, the printing press, and the new science – people began to think of themselves as self-sufficient individuals. Descartes introduced the idea of the subject as what underlay changing mental states, and Kant argued that, as the objectifier of everything, the subject must be free and autonomous. As we shall see in Chapter 4, Søren Kierkegaard concluded that each one of us is a subject called upon to take on a fixed identity that defines who one is and what is meaningful in one's world.

7 Steve Lohr, "Ideas and Trends: Net Americana; Welcome to the Internet, the First Global Colony," *The New York Times*, January 9, 2000.

8 David Blair's book, *Language and Representation in Information Retrieval*, New York, Elsevier Science, 1990, was chosen "Best Information Science Book of the Year" in 1999 by the American Society for Information Science, and Blair himself was named "Outstanding Researcher of the Year" by the same society in the same year.

9 David Blair, *Wittgenstein, Language and Information*, Springer, 2006, 287.

10 See H. Dreyfus, *What Computers (Still) Can't Do*, 3rd edn, Cambridge, MA, MIT Press, 1992.

11 See D. Lenat and R .V. Guha, *Building Large Knowledge-Based Systems*, New York, Addison Wesley, 1990.

12 Ibid.

13 V. Pratt, *CYC Report*, Stanford University, April 16, 1994.

14 D. Swanson, "Historical Note: Information Retrieval and the Future of an Illusion", *Journal of the American Society for Information Science*, vol. 32, no. 2, 1998, pp. 92–8.

15 The role of the body in our being able to experience space, time and objects is worked out in detail in S. Todes, *Body and World*, Cambridge, MA, MIT Press, 2001.

16 D. Swanson, op. cit.

17 Indeed, in spite of repeated failures, there is always some new promise on the horizon. The latest company to promise intelligent Web search is Powerset. The current hope is that syntactic natural language processing could at least allow the user's requests to be made in everyday language and, ideally, also enable the computer to understand the meaning of what is written on each Website. But the problems of capturing embodied common sense knowledge that stalled Lenat for over 20 years have not been solved. They have simply been ignored.

18 S. Brin and L. Page, "The Anatomy of a Large-Scale Hypertextual Web Search Engine," Computer Science Department, Stanford University, 1991.

19 S. Brin, R. Motwani, L. Page, T. Winograd, *What can you do with a Web in your Pocket?*, Bulletin of the IEEE Computer Society Technical Committee on Data Engineering, 1998.

20 L. Page, S. Brin, R. Motwani, T. Winograd, *The PageRank Citation Ranking: Bringing Order to the Web* (1998). Stanford Digital Libraries SIDL-WP-1999-0120.

21 From "Our Search: Google technology" at http://www.google.com/technology/

22 Ibid.

23 S. Brin, R. Motwani, L. Page, T. Winograd, *What can you do with a Web in your Pocket?* op. cit.

24 S. Brin and L. Page, "The Anatomy of a Large-Scale Hypertextual Web Search Engine."

TWO HOW FAR IS DISTANCE LEARNING FROM EDUCATION?

1 T. Oppenheimer, "The Computer Delusion", *The Atlantic Monthly*, July 1997.

2 See Dreyfus and Dreyfus, *Mind over Machine*, New York, Free Press, 1988, Chapter 5.

3 It seems that this optimism is shared in China. Reuters reports on August 22, 2000: "Chinese President Jiang Zemin offered a ringing endorsement of the Internet on Monday, saying e-mail, e-commerce, distance learning and medicine would transform China."

4 "The Paula Gordon Show", broadcast on February 19, 2000, on WGUN.

5 T. Gabriel, "Computers Can Unify Campuses, But Also Drive Students Apart", *The New York Times*, November 11, 1996.

6 For more details, see Dreyfus and Dreyfus, op. cit.

7 See M. Polanyi, *Personal Knowledge*, London, Routledge & Kegan Paul, 1958.

8 Patricia Benner has described this phenomenon in *From Novice to Expert: Excellence and Power in Clinical Nursing Practice*, Menlo Park, CA, Addison-Wesley, 1984, p. 164. Furthermore, failure to take risks leads to rigidity rather than the flexibility we associate with expertise. When a risk-averse person makes an inappropriate decision and consequently finds himself in trouble, he tries to characterize his mistake by describing a certain class of dangerous situation and then makes a rule to avoid that type of situation in the future. To take an extreme example, if a driver, hastily pulling out of a parking space, is side-swiped by an oncoming car he mistakenly took to be approaching too slowly to be a danger, he may resolve to follow the rule, never pull out if there is a car approaching. Such a rigid response will make for safe driving in a certain class of cases, but it will block further skill refinement. In this case, it will prevent acquiring the skill of flexibly pulling out of parking places. In general, if one follows general rules one will not get beyond competence. Progress is only possible if, responding quite differently, the driver accepts the deeply-felt consequences of his action without detachedly asking himself what went wrong and why. If he does this, he is less likely to pull out too quickly in the future, but he has a much better chance of ultimately becoming, with enough frightening or, preferably, rewarding experiences, a flexible, skilled driver.

One might object that this account has the role of emotions reversed;

that the more the beginner is emotionally committed to learning, the better, while an expert could be, and, indeed, often should be, coldly detached and rational in his practice. This is no doubt true, but the beginner's job is to follow the rules and gain experience, and it is merely a question of motivation whether he is involved or not. Furthermore, the novice is not emotionally involved in *choosing* an action, even if he is involved in its outcome. Only at the level of competence is there an emotional investment in the *choice of action*. Then emotional involvement seems to play an essential role in switching one over from what one might roughly think of as a left-hemisphere analytic approach to a right-hemisphere holistic one. Of course, not just any emotional reaction, such as enthusiasm, or fear of making a fool of oneself, or the exultation of victory, will do. What matters is taking responsibility for one's successful and unsuccessful choices, even brooding over them; not just feeling good or bad about winning or losing, but replaying one's performance in one's mind step by step or move by move. The point, however, is not to *analyse* one's mistakes and insights, but just to *let them sink in*. Experience shows that only then will one become an expert. After one becomes an expert one can rest on one's laurels and stop this kind of obsessing, but if one is to be the kind of expert who goes on learning, one has to go on dwelling emotionally on what critical choices one has made and how they affected the outcome.

9 Mirror-neurons probably play a role in such learning by imitation. (See 31 in Chapter 5.)

10 K. Nielsen, "Musical Apprenticeship, Learning at the Academy of Music as Socially Situated", *Nordic Journal of Educational Research*, vol. 3, 1997.

11 If we take a closer look at apprenticeship, we find that this kind of training contains important insights for testing as well as teaching. The apprentice becomes a master by imitating a master. He gradually learns how to do the whole task. Since skills are not learned by components but, rather, by small holistic improvements, there is no way to test the student in each component of the relevant skill. Where mastery is at stake, the kind of examination used in most universities and necessarily on the Internet is not useful, and even counter-productive. Rather, instead of giving the apprentice periodic examinations to see if he has mastered the supposed components that are supposedly mastered by students at his stage, when it seems to the master that an apprentice has

learned his craft, he is asked to do what is normally done by an expert in his domain of expertise. For example, if he is learning to make a musical instrument, he may be asked to make, say, a violin. But, without an examination scored on a normal curve, who is to decide whether or not the apprentice has made a good violin? Only an expert can tell. So the masters gather around and play the apprentice's violin to test it. If the apprentice has made a good violin, he is sent to another master. Otherwise, he is put back to work to gain more experience.

12 To get at the gist of the way style works, I've simplified the specific socio-logical claims. For more precise details, see, for example, W. Caudill and H. Weinstein, "Maternal Care and Infant Behavior in Japan and America", *Readings in Child Behavior and Development*, in C. S. Lavatelli and F. Stendler (eds), New York, Harcourt Brace, 1972, 78.

13 Deep Blue, the program that is currently world chess champion, is not an expert system operating with rules obtained from experts. Experts look at at most 200 possible moves, while Deep Blue uses brute force to look at a billion moves a second and so can look at *all* moves seven moves into the future without needing to understand anything.

14 Yeats's last letter, written to Lady Elizabeth Pelham just before his death. *The Letters of W. B. Yeats*, ed. Allen Wade, New York, Macmillan, 1955, 1922.

THREE DISEMBODIED TELEPRESENCE AND THE REMOTENESS OF THE REAL

1 E. M. Forster, "The Machine Stops", *The New Collected Short Stories*, London, Sidgwick & Jackson, 1985. Written in 1909 partly as a rejoinder to H. G. Wells's glorification of science, "The Machine Stops" is set in the far future, when mankind has come to depend on a worldwide machine for food and housing, communications and medical care. In return, humanity has abandoned the earth's surface for a life of isolation and immobility. Each person occupies a subterranean hexagonal cell where all bodily needs are met and where faith in the Machine is the chief spiritual prop. People rarely leave their rooms or meet face-to-face; instead they interact through a global web that is part of the Machine.

2 This sense of leaving behind one's body is also experienced when one does theoretical work. Descartes tells us that, in order to write his *Meditations*, he retired into a warm room where he would be free from

passions and from having to act. Of course, one runs the risk that, from the detached, theoretical perspective, one may get a strange idea of what it is to be a human being, and, indeed, Descartes came to the conclusion that his body was not essential to him.

3 J. Mark, "Portrait of a Newer, Lonelier Crowd is Captured in an Internet Survey", *The New York Times*, February 16, 2000.

4 Ibid.

5 G. Johnson, *Wired Magazine*, January 2000.

6 Saint Augustine, *Confessions*, trans. R.S. Pine-Coffin, London, Penguin, 1961, p. 114.

7 I. Hacking, *Representing and Intervening*, Cambridge, Cambridge University Press, 1983, p. 194.

8 René Descartes, "Dioptric", *Descartes: Philosophical Writings*, ed. and trans. Norman Kemp Smith, New York, Modern Library, 1958, p. 150.

9 Ibid., p. 235.

10 Ken Goldberg's famous piece of Web art, "The Telegarden", is an example of such interaction at a distance. Visitors to this garden log in from terminals all over the world, directing a robot and camera to view, plant, and water seeds in a 6 ft. × 6 ft. patch of soil in a museum in Austria.

11 M. Merleau-Ponty, *Phenomenology of Perception*, trans. Colin Smith, London, Routledge & Kegan Paul, 1979, p. 302.

12 Ibid., p. 250.

13 This claim is argued for at length in Samuel Todes's *Body and World*, Cambridge, MA, MIT Press, 2001.

14 Merleau-Ponty, op. cit., p. 250.

15 R. M. Held and N. I. Durlach, "Telepresence", *Presence*, vol. 1, pp. 109–11, as cited in Ken Goldberg (ed.), *The Robot in the Garden: Telerobotics and Telepistemology in the Age of the Internet*, Cambridge, MA, MIT Press, 2000.

16 J. Canny and E. Paulos, "Tele-Embodiment and Shattered Presence: Reconstructing the Body for Online Interaction", in Goldberg (ed.), op. cit. (More on this in Chapter 5.)

17 Personal communication.

18 M. Heidegger, *The Fundamental Concepts of Metaphysics*, trans. W. McNeil and N. Walker, Bloomington, IN, Indiana University Press, 1995, pp. 66–7.

19 Personal communication.

20 W. H. Graves, " 'Free Trade' in Higher Education: The Meta University", *Journal of Asynchronous Learning Networks*, vol. 1, Issue 1 – March 1997.

21 Merleau-Ponty, op. cit., p. 136.

22 That is, the player's gaze can't penetrate the distance to bring out more and more detail the way it does in real life. As Merleau-Ponty puts it: "When, in a film, the camera is trained on an object and moves nearer to it to give a close-up view, we can *remember* that we are being shown the ashtray or an actor's hand, we do not actually identify it. This is because the screen has no horizons. In normal vision, on the other hand, I direct my gaze upon a sector of the landscape, which comes to life and is disclosed, while the other objects recede into the periphery and become dormant, while, however, not ceasing to be there." *Phenomenology of Perception*, 68.

23 Merleau-Ponty talks of the sense we have in the real world of there being no sharp boundary at the edge of our visual field, but, rather, of the world continuing behind our back. He further points out that if we felt the world behind us broke off suddenly, the scene in front of us would look different. "The objects behind my back are . . . not represented to me by some operation of memory or judgment; they are present, they *count* for me. . . ." *Sense and Non-Sense*, trans. Hubert L. Dreyfus and Patricia Allen Dreyfus, Northwestern University, 1964, 51.

24 Personal communication.

25 Forster, op. cit.

26 Experiments in computer-supported cooperation have shown that people are more inclined to defect in on-line communications than in face-to-face interactions, and that a preliminary direct acquaintance between people reduces this effect. So, computer technology can even weaken trust relationships already holding in human organizations and relations, and aggravate problems of deception and trust. See C. Castelfranchi and Y. H. Tan (eds), *Trust and Deception in Virtual Societies*, Springer, 2001.

27 See D. N. Stern, *The Interpersonal World of the Infant*, New York, Basic Books, 1985.

28 Yet people say MUD (Multi User Dungeon) users fall in love in their chat rooms. I don't know what to make of that. Do they really trust each other, or, does such attraction perhaps show, as Shakespeare saw, that the erotic is more verbal than physical. (See, for example, Ulysses' description of the erotic attraction of Cressida, in *Troilus and Cressida*, IV, v, ll. 35–63.)

29 H. F. Harlow and R. R. Zimmerman, "Affectional Responses in the

Infant Monkey", *Science*, v, 130, 1959, pp. 421–32, H. F. Harlow and M. H. Harlow, "Learning to Love", *American Scientist*, v, 54, 1966, pp. 244–72. In the experiment, an orphaned monkey was given two surrogate "mothers" – a wire one and a terry-cloth one. To make the wire one more appealing, Harlow made the feeding bottle part of the wire monkey. But in spite of this, whenever the small monkey was frightened, he would scurry to the terry-cloth monkey, not the wire one.

FOUR NIHILISM ON THE INFORMATION HIGHWAY

1 S. Kierkegaard, "The Present Age", *A Literary Review*, trans. A. Hannay, London/New York, Penguin, 2001.

2 Ibid., p. 59.

3 S. Kierkegaard, *Journals and Papers*, ed. and trans. H.V. Hong and E.H. Hong, Bloomington, IN, Indiana University Press, vol. 2, no. 483.

4 Ibid., no. 2163.

5 Ibid.

6 J. Habermas, *The Structural Transformation of the Public Sphere*, Cambridge, MA, MIT Press, 1989.

7 Ibid., p. 94.

8 Ibid., p. 130.

9 Ibid., pp. 131, 133.

10 Ibid., p. 138.

11 Ibid., p. 134.

12 Ibid., p. 137.

13 Kierkegaard, "The Present Age", p. 62.

14 Ibid. pp. 62, 63. (My italics.)

15 Ibid., p. 77.

16 Ibid., p. 42.

17 Ibid., p. 68. (Kierkegaard's italics.)

18 Ibid., p. 77.

19 Kierkegaard, *Journals and Papers*, vol. 2, no. 480.

20 Ibid., no. 489. Kierkegaard would no doubt have been happy to transfer this motto to the Web, for just as no individual assumes responsibility for the consequences of the information in the press, no one assumes responsibility for even the accuracy of the information on the Web. Of course, no one really cares if it is reliable, since no one is going to act on it anyway. All that matters is that everyone pass the word along by

forwarding it to others. The information has become so anonymous that no one knows or cares where it came from. Just to make sure no one can be held responsible, in the name of protecting privacy, ID codes are being developed that will assure that even the sender's address will remain secret. Kierkegaard could have been speaking of the Internet when he said of the Press: "It is frightful that someone who is no one . . . can set any error into circulation with no thought of responsibility and with the aid of this dreadful disproportioned means of communication" (Kierkegaard, *Journals and Papers*, vol. 2, no. 481).

21 Kierkegaard, "The Present Age", p. 64.

22 Although Kierkegaard does not mention it, what is striking about such interest groups is that no experience or skill is required to enter the conversation. Indeed, a serious danger of the public sphere, as illustrated on the Internet, is that it undermines expertise. As we saw in Chapter 2, acquiring a skill requires interpreting the situation as being of a sort that requires a certain action, taking that action, and learning from the results. As Kierkegaard understood, there is no way to gain practical wisdom other than by taking risky action and thereby experiencing both success and failure. Otherwise, the learner will be stuck at the level of competence and never achieve mastery. Thus the heroes of the public sphere who appear on serious radio and TV programmes, have a view on every issue, and can justify their view by appealing to abstract principles, but they do not have to act on the principles they defend and therefore lack the passionate perspective that alone can lead to egregious errors and surprising successes and so to the gradual acquisition of practical wisdom.

23 S. Kierkegaard, *Edifying Discourses*, ed. P. L. Holmer, New York, Harper Torchbooks, 1958, p. 256.

24 Ibid., p. 260.

25 Ibid., p. 262.

26 Kierkegaard, "The Present Age", p. 103.

27 Ibid., p. 79.

28 Given Kierkegaard's use of the term "sphere", then, precisely because reflection is the opposite of taking any decisive action, and therefore the opposite of making anything absolute, what Habermas calls the public sphere is not a sphere at all.

A related non-sphere worth noting because it has become popular on the Net is Teilhard de Chardin's Noosphere, which has been embraced by the Extropians and others who hope that, thanks to the World Wide Web, our minds will one day leave behind our bodies. The Noosphere or mind sphere (in Ionian Greek "noos" means "mind") is supposed to be the convergence of all human beings in a single giant mental network that would surround the Earth to control the planet's resources and shepherd a world of unified Love. According to Teilhard, this would be the Omega or End-Point of time.

From Kierkegaard's perspective, the Noosphere, where risky, embodied locality and individual commitment would have been replaced by safe and detached ubiquitous contemplation and love, would be a confused Christian version of the public sphere.

29 S. Turkle, *Life on the Screen: Identity in the Age of the Internet*, New York, Simon and Schuster, 1995, pp. 263–4.

30 Ibid., p. 180.

31 Ibid., p. 26.

32 A year after the publication of her book, Turkle seems to have had doubts about the value of such experiments. She notes that: "Many of the people I have interviewed claim that virtual gender-swapping (pretending to be the opposite sex on the Internet) enables them to understand what it's like to be a person of the other gender, and I have no doubt that this is true, at least in part. But as I have listened to this boast, my mind has often travelled to my own experiences of living in a woman's body. These include worry about physical vulnerability, fears of unwanted pregnancy and infertility, fine-tuned decisions about how much make-up to wear to a job interview, and the difficulty of giving a professional seminar while doubled over with monthly cramps. Some knowledge is inherently experiential, dependent on physical sensations" (S. Turkle, "Virtuality and its Discontents: Searching for Community in Cyberspace", *The American Prospect*, no. 24, Winter 1996).

33 Kierkegaard, "The Present Age", p. 68.

34 S. Kierkegaard, *Either/Or*, trans. D. F. Swenson and L. M. Swenson, Princeton, Princeton University Press, 1959, vol. II, pp. 16–17.

35 Ibid., vol. I, p. 46.

36 Ibid., vol. II, p. 197.

37 When I typed in Søren Kierkegaard, Google found 2,630,000 entries.

38 Kierkegaard, *Either/Or*, vol. II, p. 228.

39 J.-P. Sartre develops the idea of the absurdity of fully free choice in *Being and Nothingness*.

40 Sartre gives the example in *Being and Nothingness* of a gambler who, having freely decided in the evening that he will gamble no more, must, the next morning, freely decide whether to abide by his previous decision.

41 S. Kierkegaard, *The Sickness unto Death, A Christian Psychological Exposition for Edification and Awakening*, trans. A. Hannay, London/New York, Penguin, 1989, p. 100.

42 For Kierkegaard there are two forms of Christianity. One is Platonic and disembodied. It is expressed best in St Augustine. It amounts to giving up the hope of fulfilling one's desires in this life, and trusting in God to take care of one. Kierkegaard calls this Religiousness A, and says it is not the true meaning of Christianity. True Christianity, or Religiousness B, for Kierkegaard, is based on the Incarnation and consists in making an unconditional commitment to something finite, and having the faith-given courage to take the risks required by such a commitment. Such a committed life gives one a meaningful life in this world.

43 Ken Goldberg's telerobotic art project: *Legal Tender www.counterfeit.org* was an attempt at inducing a sense of on-line risk. Remote viewers were presented with a pair of purportedly authentic US $100 bills. After registering for a password sent to their e-mail address, participants were offered the opportunity to "experiment" with the bills by burning or puncturing them at an on-line telerobotic laboratory. After choosing an experiment, participants were reminded that it is a Federal crime to knowingly deface US currency, punishable by up to six months in prison. If, in spite of the threat of incarceration, participants click a button indicating that they "accept responsibility", the remote experiment is performed and the results are shown. Finally, participants were asked if they believed the bills and the experiment were real. Almost all responded in the negative. So they either never believed the bills were real or else they were setting up an alibi if they were accused of defacing the bills. In either case, they hadn't experienced any risk and taken any responsibility after all.

44 As Turkle puts it: "Instead of solving real problems – both personal and social – many of us appear to be choosing to invest ourselves in unreal places. Women and men tell me that the rooms and mazes on MUDs are

safer than city streets, virtual sex is safer than sex anywhere, MUD friendships are more intense than real ones, and when things don't work out you can always leave" (S. Turkle, "Virtuality and its Discontents: Searching for Community in Cyberspace", *The American Prospect*, no. 24, Winter 1996).

45 Kierkegaard, "The Present Age", p. 80.

FIVE VIRTUAL EMBODIMENT: MYTHS OF MEANING IN *SECOND LIFE*

1 This information was furnished January 30, 2008 by Peter Gray, the Public Relations person at Linden Lab, the creators of *Second Life*.

2 "If there are people who express themselves through *Second Life* metaphors, and some of them also express needs of a spiritual nature, then maybe we should not ignore the possibility to respond to their demands," explained a recent article published in the Italian Jesuits' magazine, *Civilta Cattolica*. "In fact, the digital World can be, itself, considered as 'missionary Land.' See, 'Second Life, place of worship and sometimes 'missionary land'.'" Ruth Gledhill, Religion Correspondent, *The Times*, July 30, 2007.

3 Michael Rymaszewski et al., *second life: the official guide* (Indianapolis, IN: Wiley Publishing, 2007), 3. Page numbers henceforth will be cited in parenthesis in the text.

4 Frank Rose, "How Madison Avenue Is Wasting Millions on a Deserted Second Life," *Wired Magazine*, Issue 15.08, 2007.

5 Neal Stephenson, *Snow Crash*, Bantam paperback edition, 1993.

6 Edward Castronova, *Synthetic Worlds: The business and culture of online games*, Chicago University Press, paperback edition, 2006, 276.

7 Just what the sacred was for people like the Homeric Greeks and what it could be in our world is a complicated question. Sean Kelly and I are writing a book on the subject tentatively entitled, *Luring Back the Gods: Nihilism, Fanaticism, and the Sacred in our Secular Age*. The book is set to be published roughly two years from now by Free Press. Meanwhile, the first draft, as it were, is available as Philosophy 6, "From gods to God and back," as a podcast on iTunesU or on the University of California webcast page: http://webcast.berkeley.edu/ for courses for Spring 2007.

8 Buildings, sculptors, and clothes designed and programmed in *Second*

Life aren't made of real materials and so don't have to obey the laws of physics and chemistry.

9 The exceptions are Ian Hacking's discussion in *Rewriting the Soul*, and Robert Nozick's discussion of "the experience machine" in *Anarchy, State, and Utopia*.

10 The holodeck is a simulated reality facility located on board starships in the Star Trek universe.

11 Blaise Pascal, *Pensées*, New York: E.P. Dutton & Co., 1958, 41.

12 Ibid., 49.

13 Ibid., 41. Perhaps, referring obliquely to Descartes, his nemesis and the dominant thinker of his day, Pascal adds, "Others sweat in their rooms to show to the learned that they have solved a problem in algebra, which no one had hitherto been able to solve."

14 David Pogue, "An Experiment in Virtual Living," an interview with Rosedale, *New York Times*, February 22, 2007. [My worries in brackets.]

15 Friedrich Nietzsche, *The Gay Science*, trans. Walter Kaufmann, New York: Vintage Books, 1974, 331.

16 Ibid., 283.

17 Ibid., 231.

18 Nat Goldhaber, "Where are you? Where is your body?", *New Media Magazine*, 1999.

19 Personal communication.

20 Interestingly, Stephenson prefers the real world to any metaverse. He says he has never entered *Second Life* and has requested that *Second Life* make clear that he has no affiliation with their virtual world. "I have nothing negative to say about it," Stephenson said. "[But] there are lots of unread books on my shelves and many interesting parts of the real world I haven't visited yet. Every hour I spend in a virtual reality is an hour I'm not spending reading Dickens or visiting Tuscany." See Robert K. Elder, staff reporter, "Authors foresee future as fact catches up with fiction:[Chicagoland Final, CN Edition]" *Chicago Tribune*, Nov 13, 2006, 1.12.

21 Thinking in this vein, Nietzsche says, "Convictions are prisons." Friedrich Nietzsche, *Twilight of the Idols and The Anti-Christ*, trans. R.J. Hollingdale, Penguin, 1978 edition.

22 Albert Borgmann, *Technology and the Character of Contemporary Life: A Philosophical Inquiry*, Chicago: University of Chicago Press, 1984.

23 See Maurice Merleau-Ponty, *Phenomenology of Perception*, trans. C. Smith, London and Henley: Routledge & Kegan Paul, 1981.

"A baby of fifteen months opens its mouth if I playfully take one of its fingers between my teeth and pretend to bite it. And yet it has scarcely looked at its face in a glass, and its teeth are not in any case like mine. . . . It perceives its intentions in its body, and my body with its own, and thereby my intentions in its own body" (352).

"[T]he body image ensures the immediate correspondence of what he sees done and he himself does"(354). See also, Maurice Merleau-Ponty, *Nature: Course Notes from the Collège de France*, trans. Robert Vallier, Evanston, IL: Northwestern University Press, 2003, 303. Merleau-Ponty calls this phenomenon *intercorporiality*.

24 See "The Unbearable Likeness of Being Digital: The Persistence of Non-verbal Social Norms in Online Virtual Environments," Nick Yee, Jeremy N. Bailenson, Mark Urbanek, Department of Communication, Stanford University, Francis Chang, Department of Computer Science, Portland State University, Dan Merget, Department of Computer Science, Stanford University. (*CyberPsychology and Behavior*, Vol. 1, No. 1, (2007), pp. 115–21.)

25 Although we can set up conditions that make such an event more likely as at a political rally or at a rock concert.

26 Borgmann, *Crossing the Postmodern Divide*, University of Chicago Press, 1992, 135.

27 *Snow Crash*, 64.

28 Ibid.

29 Merleau-Ponty, *Phenomenology of Perception*, 185.

30 Sandra Blakeslee, "Cells That Read Minds" *New York Times*, January 10, 2006.

31 Vittorio Gallese, *Journal of Consciousness Studies*, 8, No. 5–7, 2001, 38, 39. Gallese adds: "This implicit, automatic, and unconscious process of motor simulation enables the observer to use his/her own resources to penetrate the world of the other without the need for theorizing about it . . . A process of action simulation automatically establishes a direct implicit link between agent and observer" (44).

32 Ibid., 38, 39.

33 Martin Heidegger, *The Fundamental Concepts of Metaphysics*, Bloomington, IN: Indiana University Press, 1995, 66–67.

34 Ibid.

35 Heidegger recognizes that direct communication fails when, for example, someone who is depressed finds that no one shares his mood and concludes that it is, indeed, private, and may not even be shareable. According to Heidegger, the Cartesian misunderstanding of the communication of moods as indirect is based on such breakdown cases and misses the way everyday moods are normally communicated.

36 Ibid., 68. Thus we need to distinguish two very different functions of moods. There is (1) the sort of mood we are all in all the time and don't normally notice, but which forms the background on the basis of which local worlds come and go; and (2) moods that can and must be shared and sensed as shared to produce a local world.

37 Heidegger notes that the special power of moods is that we can't control them; rather, they govern our actions by drawing us in. He claims that for this reason the Homeric Greeks, who understood the power of moods better than we do, thought of them as being produced by gods. Different gods had different spheres of influence in which each established his or her mood. Aphrodite's special sphere of influence was, of course, the erotic. She could establish a mood in which all that mattered to those involved were the erotic possibilities of the situation. Other gods set other moods. For example, Ares set an aggressive mood in which all that mattered was ferocious fighting. Homer saw that being attuned to a situation in a particular way opens us to what matters, and the way things and people matter is then directly shared and acted upon.

In his lecture course, *Parmenides* (Martin Heidegger, *Parmenides*, trans. A. Schuwer & R. Rojcewicz, Bloomington, & Indianapolis: Indiana University Press, 1992) 106 & 111, Heidegger says: " '[A]ffective dispositions' are not to be understood in the modern subjective sense as 'psychic states.' [W]e are thinking the essence of the Greek gods more originally if we call them the attuning ones. [T]hey determine every essential affective disposition from respect and joy to mourning and terror."

38 In some computer games there is a mood set by the programmers – aggressivity, for example – that all players share. But being in such a fixed mood in the real world would be pathological since normally moods change as they adapt us to changing situations.

39 Devoted denizens of synthetic worlds tell me that players in these

worlds share intense moods such as fear when, for example, they band together to kill a dragon in a creepy cavern. No doubt this is true. It requires us to distinguish (1) *contagious* moods *contributed to by the participants* in an enclosed focal event such as a marriage ceremony or a graduation, and (2) the sort of mood *produced by the setting* as in a cathedral, or, in this example, a dragon's lair. One can capture the difference by distinguishing the mood in a room or environment, such as restless, gay, solemn, and so forth, from the mood of a room or environment such as warm, frightening, restful, and so forth.

40 Personal communication. Nov. 2007.

41 The limited expressivity of current avatar faces and bodies is not encouraging. Perhaps, however, the technology of motion-capture can circumvent this problem.

42 This raindrop metaphor is meant to capture the contagion of moods. It might, however, be misleading in that it doesn't distinguish being swept *away* by a mood so that one *loses oneself* from the sense that, as one is swept *up* by the mood, one is also *contributing to* the character and power of the mood. There are, then, two ways of being drawn in by a mood. One can be taken over by say a mob or a fascist political rally in which one senses the overwhelming power of the crowd and the *total loss of one's freedom*, or one can feel *empowered* by the sense that one is contributing to the shared joy, sorrow, good vibes, or the like of a focal occasion.

Besides the question to what extent the mood in a room can be directly shared, another empirical question is to what extent in *Second Life* one can experience the mood of a room. Architects are sensitive to the mood of a setting – a building, a house, a room, and so forth. They would like to be able to build a virtual room in *Second Life*, have someone walk their avatar through it, and then find out whether the room felt warm, soothing, exciting, depressing, etc. (This is an entirely different question from whether there is a shared mood in the room. The room could itself be soothing but nonetheless, on some occasion, it could be the locus of an excited group mood.)

Intercorporiality is not relevant in the case of the mood of a space since the space can have its effect on a single person. But architects could experiment with the effect of various virtual spaces on the user's mood. This raises the question of the reliability of the experience of the virtual room as a predictor of what the experience of the actual room would be. I suspect that unless the person whose avatar is exploring the room takes the first person point of view, and what's more, senses his body as moving through the room, the experience of the mood of the room in *Second Life* will not be a reliable predictor of the mood of an identical actual room. But if the virtual first person perspective of walking through a room were improved so that the user had a sense of directly moving her body, the mood of the virtual room might well come to resemble the mood of a similar room in the real world and so give guidance to the architect.

43 It might seem that programmers have already shown that users can be drawn in and come to share a focal event, since there exists a screen shot of a virtual funeral for a Korean woman who died after 48 hours of continuously playing *World of Warcraft*. It is revealing that the residents of virtual worlds sense the need for memorable events and attempt to set up focal occasions such as weddings and funerals. But how could the uniqueness of the specific attunement of such an event and the sense that that mood was shared be conveyed by the stereotypical gestures selected and commanded by the mourners? It certainly wouldn't be sufficient at a funeral for all involved to display their generic mourning gesture. That would not show that the mourners were experiencing a shared public mood but only that each had privately chosen to act as a mourner.

It is surely possible to convey a shared mood in a conference call and even convey the sense that it is shared. A CEO might be on the line and convey her anxiety to the group. But could one have a conference-call-funeral? That is an empirical question. Perhaps one could, but the heaviness of the mood of mourning might well require intercorporiality.

CONCLUSION

1 L. Guernsey, "The Search Engine as Cyborg", *The New York Times*, June 29, 2000.

2 There is in fact with Google a kind of symbiosis between syntactic and semantic search. Those who use Google search are not restricted to hyperlinks but bring to bear their human understanding. In other words, people see what comes back from query Q, then issue Q' and see what comes back, and then issue Q", etc. This understanding of relevance taking advantage of the ranked pages found by Google enables

searchers to see possibly relevant pages and then, if the result is not fully satisfactory, to use their common sense to further fine-tune their query. The result is the surprising success of syntactic search.

3 Personal communication. Chart 1 shows, in percent of pages viewed daily how much Wikipedia is actually gaining on Google. Source: http:/ /www.alexa.com/data/details/traffic_details/wikipedia.com?site0= google.com&site1=wikipedia.com&y=r&z=3&h=400&w=700&range =max&size=Large

4 "Internet Opens Elite Colleges to All," Justin Popoe, Associated Press, December 29, 2007. Printed by *The Washington Post*, and *The Herald Tribune*, among others.

5 Ibid.

6 "The iPod Lecture Circuit" Michelle Quin, *Los Angeles Times*, November 24, 2007.

7 ABC World News with Charlie Gibson, Saturday March 22, 2008.

8 Michelle Quinn.

9 K. Chang, "Science Times", *The New York Times*, September 12, 2000. It is typical of the field that the situation hasn't changed in the decade since that report.

10 C. Thompson, "Being There", *Fortune Magazine*, Special Issue on the Future of the Internet, 142: 8, October 2000, p. 236.

11 National Public Radio, *Talk of the Nation*, February 29, 2000.

12 H. Rheingold, *The Virtual Community: Homesteading on the Electronic Frontier*, rev. edn, Cambridge, MA, MIT Press, 2000.

13 Ibid., pp. 375, 376.

14 Ibid.

15 Ibid., p. 379.

16 Ibid., p. 382.

17 Ibid., p. 384.

18 Ibid.

19 Andrea L. Foster, "Professor Avatar", *The Chronicle of Higher Education: Information Technology*, September 21, 2007.

20 This might well be a case of the Hawthorn effect – a famous study that showed that changing the light in a factory from incandescent to fluorescent increased productivity, but so did changing it back.

21 F. Nietzsche, *Thus Spake Zarathustra*, trans. W. Kaufmann, New York, Viking Press, 1966, p. 34.

Index

Related titles from Routledge

On Waiting
Harold Schweizer

'This is a quite remarkable book, a pleasure to read. Not only is it clear and informative but also by turns witty, melancholic and insightful. The book is astonishingly erudite, but wears this learning so lightly and so charmingly that it is both easy and gripping to read.' – *Robert Eaglestone, Royal Holloway, University of London*

Penelope waits by her loom for Odysseus, Vladimir and Estragon wait for Godot, all of us have to wait: for buses, phone calls and the kettle to boil. But do we know what the checking of one's watch and pacing back and forth is really all about? What is the relationship between waiting and time? Is there an ethics of waiting, or even an art of waiting? Do the Internet, online shopping, and text messaging mean that waiting has come to an end?

On Waiting explores such and similar questions in compelling fashion. Drawing on some fascinating examples; from the philosopher Henri Bergson's musings on a lump of sugar, to Kate Croy waiting in *Wings of the Dove* to the writings of Rilke, Bishop, and Carver, *On Waiting* examines this ever-present yet overlooked phenomenon from diverse angles in fascinating style. On Waiting is the first book to present a philosophy of waiting.

ISBN13: 978–0–415–77506–9 (hbk)
ISBN13: 978–0–415–77507–6 (pbk)
ISBN13: 978–0–203–92715–1 (ebk)

Related titles from Routledge

On Education
Harry Brighouse

'a rare example of a philosophical discourse with a direct relevance to contemporary policymaking...If forthcoming debates about education policy do not draw heavily on what he has to say here, then they will be severely impoverished' – *Julian Baggini, Times Educational Supplement*

'Clearheaded, acutely perceptive, and utterly lucid, this is the one book about education which everyone can and should make time to read.' – *Randall Curren, University of Rochester, USA*

'This is a clearly structured and thought-out book...It's polemical but also introduces the reader to key arguments and issues.' – *Stephen Law, Royal Institute of Philosophy*

What is education for? Should it produce workers or educate future citizens? Is there a place for faith schools – and should patriotism be taught?

In this compelling and controversial book, Harry Brighouse takes on all these urgent questions and more. He argues that children share four fundamental interests: the ability to make their own judgements about what values to adopt; acquiring the skills that will enable them to become economically self-sufficient as adults; being exposed to a range of activities and experiences that will enable them to flourish in their personal lives; and developing a sense of justice.

He criticises sharply those who place the interests of the economy before those of children, and assesses the arguments for and against the controversial issues of faith schools and the teaching of patriotism.

Clearly argued but provocative, *On Education* draws on recent examples from Britain and North America as well as famous thinkers on education such as Aristotle and John Locke. It is essential reading for anyone interested in the present state of education and its future.

ISBN 13: 978–0–415–32789–3 (hbk)
ISBN 13: 978–0–415–32790–9 (pbk)
ISBN 13: 978–0–203–39074–0 (ebk)

Available at all good bookshops
For ordering and further information please visit:

www.routledge.com

Related titles from Routledge

On Religion
John D. Caputo

'Intellectual without being overly academic...one cheers his vigor and relishes his insights into the paradoxical, ambiguous nature of religion and religious belief. Recommended.' – *Library Journal (US)*

'With some deft sophistry (heavily influenced by Derrida who also produced one of the other five books in the Routledge's new Thinking in Action series) John D. Caputo redefines religion as love of the unforeseeable. And, as that is a given in life, his definition of religiosity pretty much equates with my definition of joie de vivre. So the opposite of a religious person is not an atheist, merely a "pusillanimous curmudgeon". But it's not all just clever wordplay. With his unorthodox definitions in place, Caputo goes on to denounce dogma, put Marx, Nietzsche and Freud in their historical places and to reunite religion, mysticism and science. On top of all that, there's a detailed deconstruction of religion in Star Wars. I'm converted.' – *Laurence Phelan, Independent on Sunday*

'I feel obliged to warn readers that I loved this book. I loved its passion, loved its ideas, and the loved the alternately sassy and incantatory rhythms of its prose ... get this book and read it' – *Sea of Faith*

On Religion is a thrilling and accessible exploration of religious faith today. If God is dead, why is religion back? Digging up the roots of all things religious, John D. Caputo inspects them with clarity and style.

Along the way, some fascinating questions crop up: What do I love when I love my God? What can the film *Star Wars* tell us about religion, and what does "may the force be with you" really mean? What are people doing when they perform an act "in the name of God"?

ISBN13: 978–0–415–23332–3 (hbk)
ISBN13: 978–0–415–23333–0 (pbk)

Available at all good bookshops
For ordering and further information please visit:
www.routledge.com

Related titles from Routledge

On Criticism

Noël Carroll

"This book is badly needed, as much by critics as those who read them, as much by teachers of criticism as those who would like to write criticism." – *Arthur C. Danto*

"This little book runs directly counter to the modern orthodoxy that proper art criticism is all about interpretation and contextualizing. With admirable clarity and disarming candor, it defends the unfashionable view that the heart of art criticism is giving reasoned evaluations of artistic achievement. Everything else passing under this label – from gender theory to Derridean deconstruction – is secondary. What makes the book especially persuasive is Noël Carroll's unrivalled expertise in all things aesthetic." – *Gordon Graham, Princeton University*

In a recent poll of practicing art critics, 75 percent reported that rendering judgments on artworks was the least significant aspect of their job. This is a troubling statistic for philosopher and critic Noël Carroll, who argues that that the proper task of the critic is not simply to describe, or to uncover hidden meanings or agendas, but ultimately to determine what is of value in art.

Carroll argues for a humanistic conception of criticism which focuses on what the artist has achieved by creating or performing the work. Whilst a good critic should not neglect to contextualize and offer interpretations of a work of art, he argues that too much recent criticism has ignored the fundamental role of the artist's intentions.

Including examples from visual, performance and literary arts, and the work of contemporary critics, Carroll provides a charming, erudite, and persuasive argument that appraisal and evaluation of art are an indispensable part of the conversation of life.

Noël Carroll, is the Andrew W. Mellon Professor of the Humanities at Temple University, USA.

ISBN13: 978-0-415-39620-2 (hbk)
ISBN13: 978-0-415-39621-9 (pbk)

Available at all good bookshops
For ordering and further information please visit:

www.routledge.com

Related titles from Routledge

Studying Using the Web: The Student's Guide to Using the Ultimate Information Resource

Brian Clegg

Everyone uses the Internet in their school work – but if you aren't careful you can end up plodding around the information fast track. It's time to change up to something more powerful. Knowing how to get the best out of the web doesn't just make it easier and quicker to find the right information. It can also transform your school or college work into something original and outstanding.

Anyone can type a few keywords into a search engine. But that's only the beginning. With *Studying Using the Web* you can find the right material, check its authenticity, transform it into your own original work, and keep up-to-date on essential topics.

There are plenty of guide books that will point your way to interesting websites. They're great, but they get out of date very quickly, and they won't help you make something of what you find. *Studying Using the Web* is different. It's about how to find the right information, and how to make the most of it.

Studying Using the Web shows you how to:

- Know what to look for;
- Make the best use of search facilities;
- Gather pictures, sounds and more;
- Make use of the human side of the internet;
- Learn how to test information with a trust CSI kit;
- Collect and structure your information effectively;
- Make text your own; and
- Keep up to date!

You could stay jogging round the information track. But think how much better you could do with the right technology and skills to harness a leading edge study machine. Move into the study fast track now.

ISBN 13: 978–0–415–40372–6 (hbk)
ISBN 13: 978–0–415–40374–0 (pbk)
ISBN 13: 978–0–203–96709–6 (ebk)

Available at all good bookshops
For ordering and further information please visit:

www.routledge.com

Related titles from Routledge

On Landscapes
Susan Herrington

"The strength of *On Landscapes* lies in its accessibility and the interesting range of projects chosen from a diverse selection of historical periods and cultural locations. The writing style is lucid and wonderfully rich, which makes for a really engaging as well as visually stimulating book." – *Jane Rendell, Bartlett School of Architecture, University College London*

"*On Landscapes* takes a fresh, contemporary look at the ever-important topic of landscapes. Herrington mixes impressive knowledge and expertise with a clear, accessible style, using a range of cutting edge examples from everyday landscapes to grand gardens. The coverage is admirable, and the author's discussion of both historical and contemporary ideas provides useful knowledge and context in the evolution of this type of environment." – *Emily Brady, University of Edinburgh*

There is no escaping landscape: it's everywhere and part of everyone's life. Landscapes have received much less attention in aesthetics than those arts we can choose to ignore, such as painting or music – but they can tell us a lot about the ethical and aesthetic values of the societies that produce them.

Drawing examples from a wide range of landscapes from around the world and throughout history, Susan Herrington considers the ways landscapes can affect our emotions, our imaginations, and our understanding of the passage of time. *On Landscapes* reveals the design work involved in even the most naturalistic of landscapes, and the ways in which contemporary landscapes are turning the challenges of the industrial past into opportunities for the future. Inviting us to thoughtfully see and experience the landscapes that we encounter in our daily lives, *On Landscapes* demonstrates that art is all around us.

Susan Herrington is an associate professor of landscape, architecture, and environmental design at the University of British Columbia, Vancouver, Canada.

ISBN 13: 978–0–415–99124–7 (hbk)
ISBN 13: 978–0–415–99125–4 (pbk)

Available at all good bookshops
For ordering and further information please visit:
www.routledge.com

Related titles from Routledge

On Literature
J. Hillis Miller

Debates rage over what kind of literature we should read, what is good and bad literature, and whether in the global, digital age, literature even has a future. But what exactly is literature? *Why* should we read literature? *How* do we read literature?

These are some of the important questions J. Hillis Miller answers in this beautifully written and passionate book. He begins by asking what literature is, arguing that the answer lies in literature's ability to create an imaginary world simply with words.

On Literature also asks the crucial question of why literature has such authority over us. Returning to Plato, Aristotle, and the Bible, Miller argues we should continue to read literature because it is part of our basic human need to create imaginary worlds and to have stories. Above all, *On Literature* is a plea that we continue to read and care about literature.

J. Hillis Miller is a Distinguished Research Professor at the Department of English and Comparative Literature at the University of California, Irvine, USA.

ISBN 13: 978–0–415–26124–1 (hbk)
ISBN 13: 978–0–415–26125–8 (pbk)
ISBN 13: 978–0–203–16562–1 (ebk)

Available at all good bookshops
For ordering and further information please visit:
www.routledge.com

Related titles from Routledge

The Routledge Companion to Twentieth Century Philosophy

Edited by Dermot Moran

The Twentieth century was one of the most significant and exciting periods ever witnessed in philosophy, characterised by intellectual change and development on a massive scale.

The Routledge Companion to Twentieth Century Philosophy is an outstanding and authoritative survey and assessment of the century as a whole. Written by leading international scholars such as Paul Guyer, Dan Zahavi, and Matt Matravers, and featuring twenty-two chapters, the essays in this collection examine and assess the central topics, themes, and philosophers of the Twentieth century, presenting a comprehensive picture of the period for the first time.

Featuring annotated further reading and a comprehensive glossary, *The Routledge Companion to Twentieth Century Philosophy* is indispensable for anyone interested in philosophy over the last one hundred years, and is suitable for both experts and novices alike.

ISBN 13: 978–0–415–29936–7 (hbk)

Available at all good bookshops
For ordering and further information please visit:

www.routledge.com

Related titles from Routledge

On Translation
Paul Ricoeur

'One of the most distinguished and prolific philosophers of his generation.'
– *The Daily Telegraph*

Paul Ricoeur was one of the most important philosophers of the Twentieth century. In this short and accessible book, he turns to a topic at the heart of much of his work: what is translation and why is it so important?

Reminding us that the Bible, the Koran, the Torah, and the works of the great philosophers are often only ever read in translation, Ricoeur reminds us that translation not only spreads knowledge but can change its very meaning. In spite of these risks, he argues that in a climate of ethnic and religious conflict, the art and ethics of translation are invaluable.

Drawing on interesting examples such as the translation of early Greek philosophy during the Renaissance, the poetry of Paul Celan and the work of Hannah Arendt, he reflects not only on the challenges of translating one language into another but how one community speaks to another. Throughout, Ricoeur shows how to move through life is to navigate a world that requires translation itself.

Paul Ricoeur died in 2005. He was one of the great contemporary French philosophers and a leading figure in hermeneutics, psychoanalytic thought, literary theory and religion. His many books include *Freud and Philosophy* and *Time and Narrative*.

ISBN 13: 978–0–415–35777–7 (hbk)
ISBN 13: 978–0–415–35778–4 (pbk)

Available at all good bookshops
For ordering and further information please visit:
www.routledge.com

Related titles from Routledge

On Architecture
Fred Rush

Architecture is a philosophical puzzle. Although we spend most of our time in buildings, we rarely reflect on what they mean or how we experience them. With some notable exceptions, they have generally struggled to be taken seriously as works of art compared to painting or music and have been rather overlooked by philosophers.

In *On Architecture*, Fred Rush argues this is a consequence of neglecting the role of the body in architecture. Our encounter with a building is first and foremost a bodily one; buildings are lived-in, communal spaces and their construction reveals a lot about our relation to the environment as a whole.

Drawing on examples from architects classic and contemporary such as Le Corbusier and Frank Lloyd Wright, and exploring the significance of buildings in relation to film and music and philosophers such as Heidegger and Merleau-Ponty, Fred Rush argues that philosophical reflection on building can tell us something important about the human condition.

Fred Rush is Associate Professor of Philosophy at the University of Notre Dame, Indiana, USA. He is the editor of the *Cambridge Companion to Critical Theory*.

ISBN 13: 978–0–415–39618–9 (hbk)
ISBN 13: 978–0–415–39619–6 (pbk)

Available at all good bookshops
For ordering and further information please visit:
www.routledge.com